ABRARY
ST. MARY'S COLLEGE OF MARYLAND
ST. MARY'S CITY, MARYLAND 20686

THE ROLE OF
MATHEMATICS IN SCIENCE

NEW MATHEMATICAL LIBRARY

PUBLISHED BY

THE MATHEMATICAL ASSOCIATION OF AMERICA

Editorial Committee

W. G. Chinn, Chairman (1983–85)   Anneli Lax, Editor
City College of San Francisco   *New York University*

Basil Gordon (1983–85)   *University of California, Los Angeles*
M. M. Schiffer (1983–85)   *Stanford University*
Ted Vessey (1983–85)   *St. Olaf College*

The New Mathematical Library (NML) was begun in 1961 by the School Mathematics Study Group to make available to high school students short expository books on various topics not usually covered in the high school syllabus. In a decade the NML matured into a steadily growing series of some twenty titles of interest not only to the originally intended audience, but to college students and teachers at all levels. Previously published by Random House and L. W. Singer, the NML became a publication series of the Mathematical Association of America (MAA) in 1975. Under the auspices of the MAA the NML will continue to grow and will remain dedicated to its original and expanded purposes.

THE ROLE OF
MATHEMATICS IN SCIENCE

by

M. M. Schiffer

Stanford University

and

Leon Bowden

University of Victoria, B.C.

30

THE MATHEMATICAL ASSOCIATION
OF AMERICA

First Printing
© Copyright, 1984 by the Mathematical Association of America, (Inc.).
All rights reserved under International and Pan-American Copyright
Conventions. Published in Washington, D.C. by the
Mathematical Association of America

Library of Congress Catalog Card Number: 84-60251

Complete Set ISBN 0-88385-600-X

Vol. 30 0-88385-630-1

Manufactured in the United States of America

Note to the Reader

This book is one of a series written by professional mathematicians in order to make some important mathematical ideas interesting and understandable to a large audience of high school students and laymen. Most of the volumes in the *New Mathematical Library* cover topics not usually included in the high school curriculum; they vary in difficulty, and, even within a single book, some parts require a greater degree of concentration than others. Thus, while the readers need little technical knowledge to understand most of these books, they will have to make an intellectual effort.

If readers have encountered mathematics so far only in the classroom, they should keep in mind that a book on mathematics cannot be read quickly. Nor must they expect to understand all parts of the book on first reading. They should feel free to skip complicated parts and return to them later; often an argument will be clarified by a subsequent remark. On the other hand, sections containing thoroughly familiar material may be read very quickly.

The best way to learn mathematics is to *do* mathematics. Readers are urged to acquire the habit of reading with paper and pencil in hand; in this way mathematics will become increasingly meaningful to them.

The authors and editorial committee are interested in reactions to the books in this series, and hope that readers will write to: Anneli Lax, Editor, New Mathematical Library, NEW YORK UNIVERSITY, THE COURANT INSTITUTE OF MATHEMATICAL SCIENCES, 251 Mercer Street, New York, N. Y. 10012.

The Editors

NEW MATHEMATICAL LIBRARY

1. Numbers: Rational and Irrational *by Ivan Niven*
2. What is Calculus About? *by W. W. Sawyer*
3. An Introduction to Inequalities *by E. F. Beckenbach and R. Bellman*
4. Geometric Inequalities *by N. D. Kazarinoff*
5. The Contest Problem Book I Annual High School Mathematics Examinations 1950–1960. Compiled and with solutions *by Charles T. Salkind*
6. The Lore of Large Numbers *by P. J. Davis*
7. Uses of Infinity *by Leo Zippin*
8. Geometric Transformations I *by I. M. Yaglom, translated by A. Shields*
9. Continued Fractions *by Carl D. Olds*
10. Graphs and Their Uses *by Oystein Ore*
11. } Hungarian Problem Books I and II, Based on the Eötvös
12. } Competitions 1894–1905 and 1906–1928, *translated by E. Rapaport*
13. Episodes from the Early History of Mathematics *by A. Aaboe*
14. Groups and Their Graphs *by I. Grossman and W. Magnus*
15. The Mathematics of Choice *by Ivan Niven*
16. From Pythagoras to Einstein *by K. O. Friedrichs*
17. The Contest Problem Book II Annual High School Mathematics Examinations 1961–1965. Compiled and with solutions *by Charles T. Salkind*
18. First Concepts of Topology *by W. G. Chinn and N. E. Steenrod*
19. Geometry Revisited *by H. S. M. Coxeter and S. L. Greitzer*
20. Invitation to Number Theory *by Oystein Ore*
21. Geometric Transformations II *by I. M. Yaglom, translated by A. Shields*
22. Elementary Cryptanalysis—A Mathematical Approach *by A. Sinkov*
23. Ingenuity in Mathematics *by Ross Honsberger*
24. Geometric Transformations III *by I. M. Yaglom, translated by A. Shenitzer*
25. The Contest Problem Book III Annual High School Mathematics Examinations 1966–1972. Compiled and with solutions *by C. T. Salkind and J. M. Earl*
26. Mathematical Methods in Science *by George Pólya*
27. International Mathematical Olympiads—1959–1977. Compiled and with solutions *by S. L. Greitzer*
28. The Mathematics of Games and Gambling *by Edward W. Packel*
29. The Contest Problem Book IV Annual High School Mathematics Examinations 1973–1982. Compiled and with solutions *by R. A. Artino, A. M. Gaglione and N. Shell*
30. The Role of Mathematics in Science *by M. M. Schiffer and L. Bowden*

Other titles in preparation

Preface

This little book is the outgrowth of a series of lectures given to a group of high school teachers and published as a mimeographed booklet by the School Mathematics Study Group. The aim of the booklet was to illustrate many ways in which mathematical methods have helped discovery in science.

The present edition has the same objective. However, we now aim at a group of readers who, we assume, are interested in mathematics beyond the level of high school mathematics. We have added material, and we occasionally use some calculus and more intricate arguments than before. We hope that we will appeal to college students and general readers with some background in mathematics. This has also led to a change in style of exposition and choice of material. If we succeed in giving an impression of the beauty and power of mathematical reasoning in science, the purpose of our work will have been achieved.

We thank Professor R. Richtmyer for his comments on our treatment of relativity, in particular for his illuminating remark that the Lorentz transformation in space gives us more physical insight than that in one dimension. This leads to a simplification in deriving the conservaton laws of mechanics, which was elegantly done and woven into Section 7.10 by Professor P. D. Lax, whom we thank also for many other remarks that helped us clarify the exposition.

We are deeply indebted to Dr. Anneli Lax, the editor of this series of books. Her editorial help has been most valuable and she has amply demonstrated that a good editor is an author's best friend.

M. M. Schiffer

L. Bowden

1984

Contents

Preface vii

Introduction 1

Chapter 1 The Beginnings of Mechanics 3

    1.1 Archimedes' Law of the Lever 3
    1.2 First Application: The Centroid of a Triangle 10
    1.3 Second Application: The Area Under a Parabola 12
    1.4 Third Application: The Law of the Crooked Lever 16
    1.5 Galileo: The Law of the Inclined Plane 18
    1.6 Stevin: The Law of the Inclined Plane 20
    1.7 Insight and Outlook 24

Chapter 2 Growth Functions 25

    2.1 The Exponential Law of Growth 25
    2.2 Maxwell's Derivation of the Law of Errors 34
    2.3 Differential and/or Functional Equations 44
    2.4 The Problem of Predicting Population Growth 45
    2.5 Cusanus' Recursive Formula for π 59
    2.6 Arithmetic and Geometric Means 69

Chapter 3 The Role of Mathematics in Optics 75

    3.1 Euclid's Optics 75
    3.2 Heron: The Shortest Path Principle 76
    3.3 Archimedes' Symmetry Proof 80

3.4	Ptolemy and Refraction	81
3.5	Kepler and Refraction	82
3.6	Fermat: The Quickest Path Principle	84
3.7	Newton's Mechanistic Theory of Light	91
3.8	Fermat Versus Newton: Experimentum Crucis	93
3.9	To Recapitulate	95
3.10	The Role of Science in Mathematics	95
3.11	Some Practical Applications of Conics	99
3.12	Conical Ingenuity; the Reflecting Telescope	100

Chapter 4 Mathematics with Matrices—Transformations — 104

4.1	Why Use Matrices?	104
4.2	Plane Analytic Geometry and Vector Addition	104
4.3	The Dot Product	107
4.4	To Relate Coordinate Geometry and Vector Algebra	108
4.5	The Law of Cosines Revisited	110
4.6	Linear Transformations of the Plane	112
4.7	Rotations	114
4.8	Composite Transformations and Inverses	115
4.9	Composition and Matrix Multiplication	117
4.10	Rotations and the Addition Formulas of Trigonometry	119
4.11	Reflections	120
4.12	Rigid Motions (Isometries)	124
4.13	Orthogonal Matrices	127
4.14	Coordinate Transformations	128
4.15	A Matter of Notation	130

Chapter 5 What is Time? Einstein's Transformation Problem — 132

5.1	The Michelson-Morley Experiment	132
5.2	What Time Is It?	136
5.3	Einstein's Space-Time Transformation Problem	138
5.4	Einstein's Solution	144
5.5	Rods Contract and Clocks Slow Down	147

Chapter 6 Relativistic Addition of Velocities — 152

6.1	Einstein's Law of Relativistic Addition	152
6.2	Rescaling Velocities	156
6.3	Experimental Verification of Einstein's Law	161
6.4	Rescaled Velocities Revisited	163

Chapter 7 Energy — **169**

- 7.1 The Two Body Impact Problem in Classical Mechanics — 170
- 7.2 The Two Body Impact Problem in the Theory of Relativity — 173
- 7.3 Admissible Energy Functions — 174
- 7.4 More About Admissible Energy Functions — 178
- 7.5 Proof that $\phi(V)$ Is Admissible — 179
- 7.6 Energy and Momentum — 181
- 7.7 The Dependence of Mass on Velocity — 186
- 7.8 Energy and Matter — 189
- 7.9 The Lorentz Transformation and the Momentum-Energy Vector — 190
- 7.10 Relativity in More than One Space Dimension — 192
- 7.11 Relativity and Electrodynamics — 196

Epilogue — **199**

Index — **201**

Introduction

For better or for worse, we are both the beneficiaries and the victims of a technological age. Every year brings new discoveries, new inventions and new technological advances, with resultant changes in the patterns of society and our modes of life. Indeed, there have been more scientific advances since 1900 than in all preceding centuries, and more advances in the last ten years than in any preceding decade. Nowadays, to be entirely innocent of science, to have no appreciation of the scientific temper, is to be increasingly out of touch with our times. Characteristically, as sciences advance they continue to become increasingly mathematical.

Let us content ourselves with just one example. Meteorology was as unscientific as it was unmathematical until it became possible to make quantitative measurements of atmospheric pressure. Until then weather forecasting was limited—with correspondingly limited success—to applications of such rough generalized observations as: red sky in the morning, shepherd's warning; red sky at night, shepherd's delight. With the invention of the barometer, the science of meteorology was born. Now that we are aware of upper air jet streams and have upper atmosphere photographs from satellites—with more mathematical physics to exploit more data—we have more success in weather prediction.

We hope that this preamble will entice readers to soldier on in pursuit of the role of mathematics in science, and that they will sense the mystery of it, the power and the glory: that the world is comprehensible; that with the hieroglyphics of mathematics, with pen and paper, we can hitch a pair of scales to a star and weigh the moon.

To appreciate the work of Archimedes, Galileo, Stevin, Newton, Maxwell, Einstein,..., to follow in the footsteps of giants, is to wear seven league boots. Fully to attain their stride is to attain their stature. To master a sustained closely reasoned mathematical argument, to grasp fully a new idea, is to acquire a brand new skill. Skills demand application and effort. It takes time for the bicyclist to learn to keep his balance, for the swimmer to feel confident in the water; it takes practice for the violinist to master his art, for the chess enthusiast to master Hamppe-Allgaier's variation of the Vienna Gambit. There is no dispensation for would-be mathematicians.

There being no dispensation, we have striven to pace our arguments so that the beginner in mathematics and science and that ubiquitous person, the general reader, can both keep in step with us without getting out of breath. Also to this end we have restricted the discussion to a level of mathematical and physical difficulty which he should find instructive yet still challenging. And rightly so. For as Herbert Spencer observed, the student will never do all that he is capable of doing if he is never required to do that which he cannot do.

The text grows in sophistication; we suggest that readers having difficulties working through the last chapters lay the book aside and return to it at a later date. Remember that in any really worthwhile endeavor a temporary setback or two is not unusual. When defeated by a sustained, complicated argument, divide to conquer. First re-read to get the general drift, then re-read again to get the detail. To do two difficult things at the same time, first do them one at a time. Also remember that great ideas are often at first difficult to recognize for what they are, though they seem deceptively simple. Only after we know their wide applicability can we appreciate their power.

CHAPTER ONE

The Beginnings of Mechanics

1.1 Archimedes' Law of the Lever

We start with the simplest machine known to mankind, the lever. Supposedly, ever since man developed beyond the level of the ape, he has used sticks to lever stones. The Egyptians in building their pyramids used elaborate machines consisting of a combination of levers; yet their knowledge of levers appears to have remained largely inarticulated. We all know that in pushing a door shut, the nearer the point at which we push it is to the line of the hinges, the harder we need push. Yet how many of us realize that this common experience exemplifies the law of the lever? The hinge is the fulcrum about which the turning moment of our push counterbalances the opposing turning moment of friction at the hinge. We have the experience, but not the articulation.

It seems that Archimedes (287–212 B.C.) was the first in history to ask for a precise mathematical formulation of the conditions of equilibrium of the lever. To ask this question was itself a tremendous step—to ask for mathematical laws for the behavior of a combination of sticks and stones; for here is a crucial novelty—*that number plays a role in understanding and predicting nature.*

We now retrace the essential steps by which Archimedes derived his formulation. He started with the simplest case: a weightless lever with equal arms suspending equal weights. See Figure 1.1.

Question: Which weight sinks? By the law of insufficient reason there is no more cause for the left-hand weight to sink than the right; by the law of sufficient reason there is as much reason for the left not to sink as for the right; the figure is symmetrical. That is, the lever does not move at all.

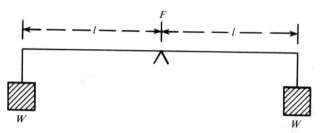

Figure 1.1

We cannot prove this by mathematics. Resort to the law of sufficient (or insufficient) reason is really an appeal to our common experience. So, with Archimedes, we take it as axiomatic that a lever as illustrated in Figure 1.1. is in equilibrium. We shall refer to this as *Archimedes' Axiom*.

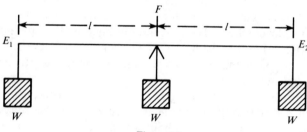

Figure 1.2

At first sight such a beginning seems too trivial to be capable of development. And yet...? Consider Figure 1.2. Although we have added a third weight to W to the weightless lever at its fulcrum, we still have symmetry, and so we still have equilibrium. We now take the two weights suspended from E_1 and F and hang both of them at M_1, the midpoint of $E_1 F$. See Figure 1.3.

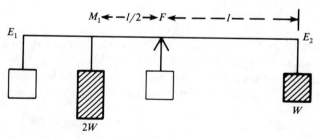

Figure 1.3

Granted that we have weightless strings, we wish to argue that we still have equilibrium. From common experience we all know that if a ladder of uniform density and weight $2W$ is supported by the shoulders of two men (of equal height), one at each end, then the ladder presses on each man with weight W. Equally well we all know that if the ladder were to be supported only by one man at its center, then he would be pressed upon by the full weight $2W$. Indeed these considerations are implicit in Figure 1.1; $2W$ at the fulcrum counterbalances W at each end of the lever.

If, in particular, a uniform ladder $E_1 F$ is supported by one man at its midpoint M_1 instead of by two men at its endpoints E_1, F, it remains in equilibrium; in neither case has it a tendency to rotate about F. And now if $E_1 F$ is given a weightless extension to E_2, surely there is no reason to suppose that its static equilibrium is being disturbed. In other words, on the basis of common experience we argue that if a weightless lever is in static equilibrium, then it will continue to be in static equilibrium when any two equal weights W suspended from it at points P_1, P_2 are replaced by a weight $2W$ suspended from the midpoint of $P_1 P_2$. We thus argue that Figure 1.3 illustrates static equilibrium. We hardly need add that since the lever itself is weightless we can discard the $E_1 M_1$ portion without disturbing equilibrium.

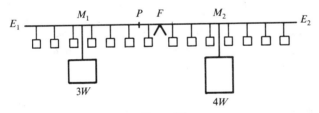

Figure 1.4

Next, suppose 14 weights, each of $\frac{1}{2}W$, to be suspended at equal intervals, say of unit length, from a weightless lever $E_1 E_2$ and to ensure equilibrium, to be symmetrically placed with respect to a fulcrum F at the midpoint of $E_1 E_2$. Now conceive the lever to be partitioned at any point P whatsoever between the points of suspension of the 6th and 7th weights, as illustrated in Figure 1.4, where M_1 is the point on the lever midway between the points of suspension of the 3rd and 4th weights, and M_2 is the point midway between the points of suspension of the 10th and 11th weights.

Mindful of the situation illustrated by Figure 1.3, we note that the 3rd and 4th weights, each equal to $\frac{1}{2}W$, may be replaced by a single weight

W at M_1 without disturbing equilibrium. Since all intervals between consecutive weights are equal, the 2nd and 5th weights are also symmetrically located with respect to M_1 and can be replaced by a single weight W suspended at M_1. Similarly, the 1st and 6th weights can be replaced by a single weight suspended from M_1. Thus the 6 weights hanging from (the weightless) E_1P can be replaced by their total weight $3W$ suspended from M_1 without disturbing equilibrium. Similarly the 8 weights suspended from (the weightless) E_2P can be replaced by their total weight $4W$ suspended at M_2 without disturbing equilibrium. In short, $3W$ at a distance 4 units from F counterbalances $4W$ at a distance 3 units from, and on the other side of, F.

In a similar vein the reader is urged to suppose P to be any point between the points of suspension of the 1st and 2nd weights and to conclude that $\frac{1}{2}W$ at a distance of $6\frac{1}{2}$ units from F counterbalances $6\frac{1}{2}W$ at a distance of $\frac{1}{2}$ unit from, and on the other side of, F. More generally he is urged to examine the consequences of placing P successively between the 2nd and 3rd points of support, the 3rd and 4th (worked out above), the 4th and 5th, ..., and to conclude that for $d = \frac{1}{2}, 1, \frac{3}{2}, 2, \ldots, 6\frac{1}{2}$, a weight of dW at a distance $7 - d$ units to the left of F counterbalances a weight of $(7-d)W$ at a distance d units to the right of F.

Figure 1.3 and variants of Figure 1.4 enable us to reduce complex cases of equilibrium to simpler, self-evident cases. With this insight Archimedes' generalization to the general law of equilibrium is readily at hand. Despite its elegance it appears that Galileo found it somewhat ponderous, and so it is to Galileo's modification that we now turn.

To follow Galileo's modification we picture a homogeneous beam of constant cross-section suspended by weightless strings from the ends of a weightless lever E_1E_2 whose fulcrum F is at its midpoint; see Figure 1.5. By considerations of symmetry there is no reason why the lever should tilt either way. We have equilibrium.

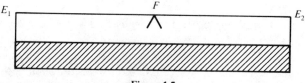

Figure 1.5

We now imagine the homogeneous beam to be cut into two beams of weights W_1, W_2 by a vertical plane through an arbitrary point P in E_1E_2. We suppose there is no loss of material in cutting the beam and consequently no loss in weight. But as we now have two beams, to prevent the one from dangling from E_1 and the other from E_2, we introduce a pair of weightless strings from P to the beams as shown in Figure 1.6

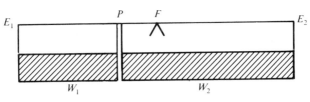

Figure 1.6

With the introduction of these two strings, there are of course two downward pulls on the lever at P. The symmetry of forces acting on the lever in Figure 1.5 has been destroyed. We have to ask: has not also equilibrium been destroyed? In Figure 1.5 the lever is in equilibrium under equal downward forces at E_1, E_2. If in Figure 1.6 there were the same downward pulls at E_1, E_2, then obviously the two additional pulls at P would destroy equilibrium of the lever.

Since there was no loss of material in cutting the original beam, its weight in Figure 1.5 is $W_1 + W_2$, equally supported at E_1, E_2. Similarly, in Figure 1.6, E_1 and a string at P equally support W_1, and a string at P and E_2 equally support W_2. That is, with supporting strings at P, there is a decrease of $\frac{1}{2}W_2$ in the pull at E_1, of $\frac{1}{2}W_1$ at E_2, and a support of $\frac{1}{2}W_2 + \frac{1}{2}W_1$ at P. Due to the introduction of downward pulls at P, the downward forces at E_1, E_2 are not the same in Figure 1.6 as in Figure 1.5 —but this is not to say that equilibrium is thereby destroyed.

The redistribution of forces between lever and beam when the beam is cut is irrelevant. Our proper concern is not with equilibrium of lever alone, but with equilibrium of the entire system of lever-cum-beam. The cut beam system in Figure 1.6 has the same mass as the uncut beam system in Figure 1.5, and moreover, the same distribution of mass relative to its single point of support F. And therefore it seems intuitively reasonable to argue that if the system has equilibrium in Figure 1.5 then also it has equilibrium in Figure 1.6.

We now re-apply Archimedes' idea, illustrated both by Figure 1.3 and Figure 1.4, that instead of two men each supporting half the weight of a ladder at each end, one man can support its entire weight at its middle. In brief, we replace the two weightless strings supporting W_1 from E_1 and P by one weightless string from the midpoint M_1 of E_1P, and similarly we replace the two weightless strings supporting W_2 from P and E_2 by one weightless string from the midpoint M_2 of PE_2. See Figure 1.7.

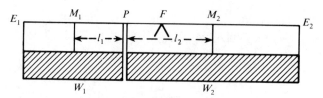

Figure 1.7

Let P be an arbitrary point on E_1E_2; set $M_1P = l_1$ and $PM_2 = l_2$. Then W_1 is the weight of a beam of length $2l_1$, W_2 that of a beam of length $2l_2$, and the overall length E_1E_2 of both beams is $2l_1 + 2l_2$. Since F is the midpoint of E_1E_2, we have $E_1F = l_1 + l_2$, so that

$$M_1F = E_1F - E_1M_1 = \tfrac{1}{2}E_1E_2 - \tfrac{1}{2}E_1P$$
$$= l_1 + l_2 - l_1$$
$$= l_2.$$

Since the beam of Figure 1.5 is homogeneous, we may without loss of generality suppose it to be of unit density. Thus we have a weight $W_1 = 2l_1$ suspended at M_1 counterbalanced by a weight $W_2 = 2l_2$ suspended at M_2. In other words, a weight $2l_1$ acting at a distance l_2 from the fulcrum is counterbalanced by a weight $2l_2$ acting at a distance l_1 from it.

What are the conditions for equilibrium? Clearly

$$2l_1 \cdot l_2 = 2l_2 \cdot l_1;$$

that is, since $W_1 = 2l_2$, $W_2 = 2l_1$,

(1.1) $$W_2 \cdot l_2 = W_1 \cdot l_1$$

or

left weight × length of left arm = right weight × length of right arm.

The product of the weight and its distance from the fulcrum is called the *moment* of the weight about the fulcrum. It is a measure of the tendency of the weight to turn the arm about the fulcrum. So, alternatively expressed, the condition for equilibrium is:

left hand moment = right hand moment.

This is Archimedes' law of equilibrium of the lever.

The illustration using 14 weights, of course, can be generalized. Let equal weights be uniformly distributed along a lever of length $2l$, and let the lever be partitioned into two segments of lengths $2a, 2b$, such that $2a + 2b = 2l$; that is, $a + b = l$. Suppose all of the weights along the left partition were stacked at the midpoint P (a units from the end) and all weights along the right partition stacked at its midpoint, Q. Then P is a distance $l - a$ from the fulcrum, and Q at distance $l - b$ ($= a$) from the fulcrum. As the weights p, q were uniformly distributed along each segment, p and q are proportional to the lengths of the respective segments: $p = ka$, $q = kb = k(l - a)$. That is,

$$k = \frac{p}{a} = \frac{q}{l-a} \quad \text{or} \quad p(l - a) = qa.$$

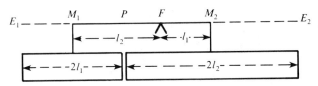

Figure 1.8a

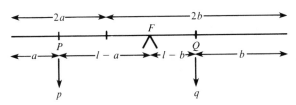

Figure 1.8b

From the figure, it is clear that $l - a$ and a are the lengths of the lever arms.

To conclude this section it is rewarding to compare Archimedes' Figure 1.4 with Galileo's Figure 1.6. Because Archimedes has, in effect, a "beam" of discrete particles instead of a beam of uniformly distributed material like Galileo, generalization is for him relatively ponderous—yet Archimedes' Figure 1.3 is surely more immediately intuitive.

We have seen how, starting from "the obvious" intuitive physics about which there is common agreement, we build up mathematics in a cumulative way. We shall now further illustrate this cumulative process by making applications of Archimedes' law.

1.2 First Application: The Centroid of a Triangle

We consider an idealized triangle, made of rigid but weightless material, lying in a horizontal plane, with a weight W suspended from each vertex. Our problem is to find the point at which the triangle can be supported without tilting from the horizontal. See Figure 1.9.

How are we to contend with three forces all at once? We must use what we know, yet the law of the lever is applicable only to two forces. In order to apply this law, we must eliminate the effect of one of the weights, say the one at A. We achieve our purpose by introducing a support at A. Now considering A', the midpoint of BC, as the fulcrum of BC, we have a lever with equal weights suspended from equal arms. Thus if the triangle is also supported at A', the points A, A', and consequently the line AA' (a median of the triangle) are fixed, so that the only motion possible is a rotation about AA'. But the forces at B, C counterbalance, so that the triangle is in equilibrium.

Obviously an upward force of W at A will counterbalance that of the weight suspended there. What upward force at A' will counterbalance the downward forces of W at B and at C? When standing on the platform of a weighing machine or scale, your weight, as indicated by the scale, is the same no matter whether you stand on one leg or both. The total

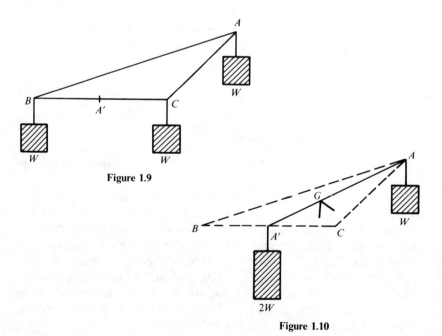

Figure 1.9

Figure 1.10

downward force is $2W$, so we require $2W$ acting upward at A'. In short, in so far as equilibrium is concerned, the original forces are equivalent to downward forces of W at A and $2W$ at A'. We have reduced a problem of three forces to a problem of two forces. See Figure 1.10.

The rest is easy, for the law of the lever is immediately applicable to this pair of forces. Let G be the point on AA' such that

$$AG = 2 \cdot A'G$$

so that

$$W \cdot AG = W \cdot (2 \cdot A'G) = 2W \cdot A'G.$$

Thus the triangle is in equilibrium when suspended at G. This solves our problem. Moreover, since the total of the downward forces at A and A' is $3W$, we conclude that the effect of the three equal forces of W at the vertices is equivalent of a force of $3W$ at G.

Articulation of our common experience and careful application of the law of the lever has solved our problem; yet we have by no means exhausted the results inherent in this problem. The argument by which we conclude that the point of suspension for equilibrium is G, two-thirds the way down the median AA', is equally applicable to the other two medians. There are no grounds for preference. Yet the forces considered can have only one resultant; consequently the three medians must be concurrent at G, a point two-thirds the way down each. See Figure 1.11.

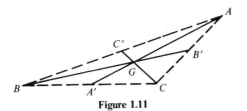

Figure 1.11

In this short deduction we see the interplay between mechanics and geometry. Not only can we use mathematics to deduce laws of nature; we can use laws of nature to deduce more mathematics. Here is an art of which Archimedes was a master.

Another result. Now suppose the horizontally placed triangle of Figure 1.11 to be a lamina of homogeneous material made up of a very large number n of equally narrow strips, each with two edges parallel to BC, as illustrated by the strip B_1C_1 in Figure 1.12. From a simple application of similar triangles it follows that any line segment parallel to BC is bisected by the median AA'. But for sufficiently large n, each hair-line fibre is bisected by, and therefore has its equilibrium point on, the median AA'. Therefore the triangular lamina in its entirety has its equilibrium point on AA'.

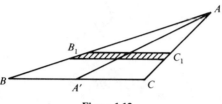

Figure 1.12

But for precisely similar reasons the equilibrium point must also lie on the other two medians, so by the foregoing result this point must be G. Thus the triangle, if horizontally oriented, can be maintained in equilibrium by a force equal to its weight acting vertically upwards at G. In short, the multitude of gravitational forces acting on the various bits of the lamina act as if they were all concentrated at G. For this reason G is known as the *center of gravity*, or *centroid*, of the triangular lamina.

1.3 Second Application: The Area Under a Parabola

Archimedes' predecessors and contemporaries had tried unsuccessfully to compute the area of an ellipse, and the area bounded by a hyperbola and one of its chords. Characteristically, Archimedes tackled the other conic section—the parabola—and was successful. His success caused a sensation, as well it might, for his method lies at the threshold of the integral calculus.

Unlike Archimedes, we have the notational conveniences afforded by analytic geometry. The problem is to find the area bounded by the parabola $y = ax^2$, the x-axis, and the vertical line $x = h$, i.e., the shaded area OAB of Figure 1.13. By symmetry it is obvious that this is one-half the area OBB_1, bounded by the line $x = h$, the given parabola, and its mirror image in OX, $y = -ax^2$. Carefully compare Figures 1.13 and 1.14. To any vertical strip PQ (of length $2ax^2$) *at a distance x from O* in Fig. 1.13 there corresponds a vertical strip $P'Q'$ (of length $2ax$) *at a distance x from O'* in Figure 1.14. As the midpoint of PQ moves from O to A, to use a favorite expression of Archimedes, PQ "fills" the area

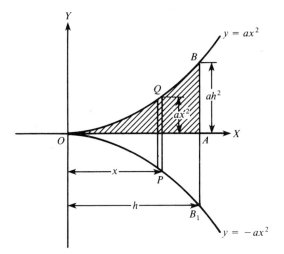

Figure 1.13

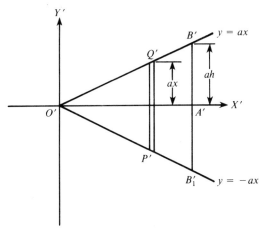

Figure 1.14

OBB_1. At the same time the midpoint of $P'Q'$ moves from O' to A' and $P'Q'$ "fills" triangle $O'B'B_1'$.

Now study the conjunction of these diagrams, see Figure 1.15; think of OB_1B as hanging in a vertical plane while triangle $O'B_1'B'$ lies flat in a horizontal plane. The lever OA' has fulcrum at O', where $OO' = 1$. We suppose the corresponding typical strips PQ, $P'Q'$ to have the same width ε, and the homogeneous material of both bodies to be of unit density. Thus the weight of the strip PQ is $2ax^2 \cdot \varepsilon \cdot 1$, and the weight of

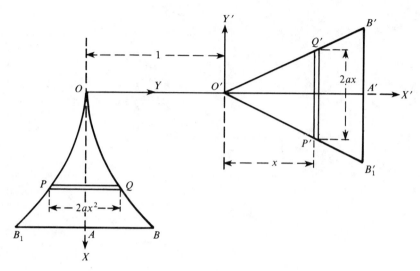

Figure 1.15

$P'Q'$ is $2ax \cdot \varepsilon \cdot 1$. The center of gravity of PQ lies vertically below O, so that the moment of PQ about O' is

$$OO' \cdot (2ax^2 \cdot \varepsilon \cdot 1) = 1 \cdot (2ax^2 \cdot \varepsilon \cdot 1) = 2ax^2\varepsilon.$$

And since $P'Q'$ is at a distance x from O', its moment about O' is

$$x \cdot (2ax \cdot \varepsilon \cdot 1) = 2ax^2\varepsilon.$$

Thus,

(1.2)   moment of PQ about O' = moment of $P'Q'$ about O',

and the corresponding strips counterbalance one another. But this result holds for each and every corresponding pair! We conclude that

moment of whole body OBB_1 about O' = moment of $OB'B'_1$ about O'.

Let W be the weight of OBB_1. Triangle $O'B'B'_1$ has altitude $O'A' = h$ and base $B'B'_1 = 2ah$ and therefore weight $\frac{1}{2} \cdot h \cdot 2ah \cdot 1 = ah^2$. This weight acts as if concentrated at G, the centroid of the triangle. By a previous result G is two-thirds the way along $O'A'$, and our law of the lever (1.1) becomes

$$W \cdot 1 = \tfrac{2}{3}h \cdot ah^2.$$

So, remembering that our materials are of unit density, we have

$$\text{Area } OBB_1 = \tfrac{2}{3}ah^3;$$

and, remembering the symmetry,

$$\text{Area } OAB \text{ under the parabola} = \tfrac{1}{3}ah^3.$$

We conclude with the elegant result that

$$\text{Area } OAB = \tfrac{1}{3}\text{Area rectangle } OABC.$$

See Figure 1.16.

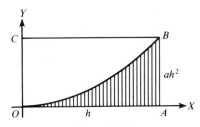

Figure 1.16

Of course this proof is not completely rigourous, since the strips PQ, $P'Q'$, of thickness ε, are not precisely rectangular. Yet it is intuitively evident that by making ε sufficiently small we make the errors of these approximations as small as we please, so that for sufficiently small ε the difference between the moments of PQ and $P'Q'$ about O' may be made arbitrarily small. Further articulation would necessitate explaining the notion of limit. To say this is not to sneer at Archimedes' proof by his "mechanical method," as he called it; on the contrary, it is to suggest that intuitive proofs are often indispensable stepping stones to more rigorous ones. Archimedes was too good a mathematician to rest content with this proof; he subsequently gave a completely rigorous one by the "method of exhaustion." The discovery (by his mechanical method) of the right formula led him to a correct proof. To cook, first catch your hare.

Archimedes' rigorous derivation of the area under the parabola, together with a dozen or so other such results, including the formula for the volume of the sphere, were known to mathematicians of the Renaissance.[†] That he had initially used a "mechanical method" was also known, but not the details. His cooking told nothing of his catching. Cavalieri

---

[†] For further details the interested reader is referred to "Episodes from the Early History of Mathematics" by A. Aaboe, New Mathematical Library, Vol. 13, (1964).

(1598–1647) devised a method for computing areas and volumes based on the intuitive consideration that if two figures have equal corresponding strips or cross sections (e. g., PQ and $P'Q'$ in Figure 1.15), then the corresponding total areas (or volumes) are equal. It was not until 1906 that a palimpsest giving the details of Archimedes' mechanical method was discovered in Istanbul, and translated by Heiberg (1854–1928), the great Danish expert on Greek mathematical texts. Had this been available to Cavalieri, his development, and consequently that of Fermat, Newton, and Leibniz would probably have been radically different.

Let us recapitulate. We began with the question, "What is the law of the lever?" Geometry, together with mechanics, enabled us to find this law and successive applications of it, reducing a problem of three forces to two, to one, determined the centroid of any triangle—and gave us, incidentally, a theorem of geometry. The notion of centroid with yet another reapplication of the law of the lever gave us the area under a parabola. This is typical of the way mathematics works: beginnings almost too trivial to take seriously, lead, with repeated application, to new insights and new discoveries, which, with further application, yield still more insight and discovery.

1.4 Third Application: The Law of the Crooked Lever

We suppose a homogeneous beam to be freely pivoted in a vertical plane about a (horizontal) nail through its geometrical center F. Suppose weights W_1, W_2 are suspended from it at A_1 and A_2 (as illustrated by Figure 1.17) so that

$$W_1 l_1 = W_2 l_2.$$

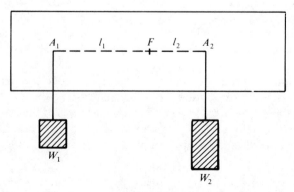

Figure 1.17

BEGINNINGS OF MECHANICS

The homogeneous beam being symmetric about F, its weight has no effect on the equilibrium of W_1, W_2; the whole figure is in equilibrium.

What changes can we make in the suspension of W_1 without disturbing equilibrium? Supposing W_1 to be a constant weight, $A_1 F$ must remain unchanged, for otherwise the turning moment of W_1 about F would be altered. But we all know that the vertical pull of a weight on its point of suspension is unchanged by shortening or lengthening the string by which it is suspended, if the string itself is of negligible weight. Clearly, W_1 may be raised or lowered; what matters for equilibrium is that its line of action, its supporting string, passes vertically through A_1.

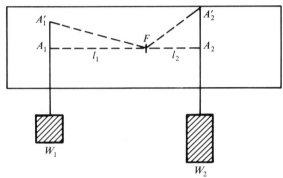

Figure 1.18

Study Figure 1.18. Does it matter that W_1, W_2 are now suspended from A'_1, A'_2, respectively, instead of from A_1, A_2? No, for the lines of actions of the two forces (and the forces themselves, of course) are unchanged.

But what is the role of the beam in this scheme of things? Being homogeneous and suspended about its geometrical center, it has no turning moment. Its total weight is counteracted by the upward pull of

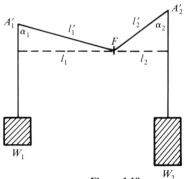

Figure 1.19

the nail; it is, in effect, weightless. Its role is to preserve the vertical lines of action $A'_1 A_1$, $A'_2 A_2$. With these preserved, the turning moments of W_1, W_2 suspended at points A'_1, A'_2 are just the same as if the points of suspension were A_1, A_2. It remains merely to idealize a little more to reject the beam's substance while preserving the lines of action. In short, equilibrium is maintained by the crooked, weightless, rigid lever $A'_1 F A'_2$ shown in Figure 1.19.

If α_1, α_2 are the angles which $A'_1 F$, $A'_2 F$ make with the vertical, we have

$$l'_1 \cdot \sin \alpha_1 = l_1, \qquad l'_2 \cdot \sin \alpha_2 = l_2$$

and, by (1.1)

(1.2′) $$W_1 l'_1 \sin \alpha_1 = W_2 l'_2 \sin \alpha_2;$$

this is the law of crooked levers. The turning moment of a force is now the product of the force, the length of the arm, *and* the sine of the angle between the line of action of the force and the arm. The factor $\sin \alpha$ is the price we pay for crookedness. Note that when $\alpha_1 = \alpha_2 = 90°$, (1.2′) becomes

$$W_1 \cdot l'_1 = W_2 \cdot l'_2.$$

Characteristically, our new result includes that from which it was deduced. Let us turn to further developments.

1.5 Galileo: The Law of the Inclined Plane

Galileo (1564–1642) was interested in the mechanics of the inclined plane. He asked and answered the question: Given a weight W on a frictionless plane inclined at an angle α to the horizontal, what force w acting up the plane is necessary to prevent W from sliding down? See Figure 1.20. Note that the precise formulation of the problem is itself a step toward solution. The unformulated, inarticulated physics of bicycling makes it obvious that the steeper the incline the greater the necessary restraining force. Clearly w is a maximum when $\alpha = 90°$, and then, since there is no help from the incline, we have $w = W$; on the other hand, if $\alpha < 90°$, then $w < W$. Thus it is appropriate to denote the restraining force by the smaller letter. We now ask the question: Can the solution of Galileo's problem be obtained as an application of the law of the crooked lever? The answer is yes—given the ingenuity of Galileo.

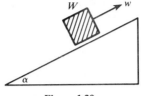
Figure 1.20

First, since vertical forces are better understood, Galileo converts w, the force along the inclined plane, into a vertical force by introducing a frictionless pulley wheel and a weightless string, as pictured in Figure 1.21. This strategem may not appear at first sight to advance the solution of the problem. But what is the problem? What weight w is needed to counterbalance W? If these are in equilibrium, there is a certain constraint between them. The connecting string being inextensible, if W moves up or down the incline a distance d, then w moves vertically down or up the same distance. Galileo had the great insight to see that this constraint could be realized in a different way—by the introduction of a crooked lever. Figure 1.22 shows an equal-armed, crooked (and rigid but weightless) lever A_1FA_2 with fulcrum F. A_1 is the center of gravity of W, and FA_1 is perpendicular to the inclined plane; A_2 is a point on the line of action of w, and FA_2 is horizontal. To satisfy ourselves that a point F satisfying these requirements exists, it is sufficient to note that the bisector of angle A_1VA_2 is the locus of points equidistant from the lines A_1V and A_2V.

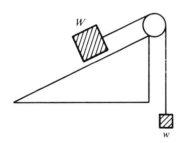
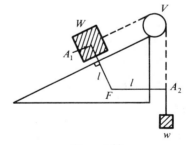

Figure 1.21   Figure 1.22

If the lever is rotated about F in a vertical plane, since the lever is equal-armed, A_1, A_2 trace out arcs of equal circles. The smaller the rotation the more nearly these arcs approximate straight lines, i.e., for infinitesimal rotations the displacement of A_1 (the center of gravity of W) along the inclined plane is the same as the vertical displacement of A_2, and therefore the same as that of w. Thus the constraint realized by string and pulley may, alternatively, be realized by the crooked lever $A_1 F A_2$. But we know the conditions for equilibrium with crooked levers, so that the problem is, in principle, solved.

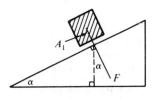

Figure 1.23

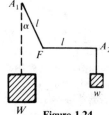

Figure 1.24

Now, the details. From Figure 1.23 it is clear that the angle between the arm $A_1 F$ and the vertical line of action of W at A_1 is α. So, by the foregoing considerations, we see that the conditions for w to maintain W in equilibrium on an inclined plane of angle α are equivalent to those for equilibrium of the crooked lever shown in Figure 1.24.

By (1.2')

$$wl \cdot \sin 90° = wl \cdot 1 = Wl \sin \alpha,$$

so that

$$w = W \sin \alpha.$$

This is the law of the inclined plane.

1.6 Stevin: The Law of the Inclined Plane

There is another proof, a most elegant proof, due to the Dutch mathematician Simon Stevin or Stevinus (1548–1620). Although Stevin was one of the most brilliant applied mathematicians who ever lived, he is less well known than Galileo; he was not a martyr of science threatened by the Inquisition. He invented the first horseless carriage, a sailing carriage for use on the dunes of the Dutch coast; he constructed famous dikes still in use today; and feeling practical need for the facility of

BEGINNINGS OF MECHANICS

decimal fractions, he invented them. For him mathematics, to be any good, had to be good for something.

Let us see how he proved the law of the inclined plane, that the force acting down it due to W is $W \sin \alpha$, where α is the angle of inclination. His proof is based on Figure 1.25. Stevin was so pleased with his proof that this diagram, with the inscription, "It looks like a miracle, but it is not a miracle," is the frontispiece of his treatise on mechanics. He had good cause for his pleasure; that the law of the inclined plane follows from the equilibrium of a heavy rope, with joined ends, when suspended over a triangular prism, was an observation of genius.

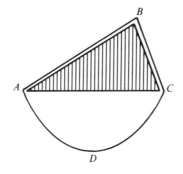

Figure 1.25

Suppose the heavy rope to be in motion initially. This assumption raises the question: when will it stop rotating? Its rotation is caused by the forces acting upon it. But for every particle of rope that goes down, say, at C, an identical particle moves up at A. Thus the configuration of the rope remains unchanged, and consequently the driving forces which initially caused motion still persist. Therefore, since it is rotating initially, it must continue to rotate forever. We have a perpetual motion machine and can use its power to drive a dynamo.

We feel, as Stevin felt, that this conclusion is absurd. But either the heavy rope is in equilibrium or it is not. With him, we have no alternative but to conclude that the rope must be in equilibrium.

Undoubtedly the portion of the rope hanging below the triangle hangs symmetrically; the downward force at A is counterbalanced by an equal downward force at C. Thus, since the rope ABC is in equilibrium before the removal of the portion ADC, it must remain in equilibrium after its removal. That is, the force acting down the one incline due to the weight of the rope BA counterbalances the force acting down the other due to the weight of the rope BC. See Figure 1.26, where AB has length a and incline α, BC has length b and incline β.

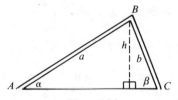

Figure 1.26

The force F necessary to prevent a weight W from sliding down an inclined plane of angle θ depends on θ, i.e. F is a function $f(\theta)$. Also, of course, F depends on W. If for a given incline W is doubled, then F is doubled; if W is trebled, the F is trebled. If W is increased k-fold, then k times the force is needed. Thus for fixed θ, we take F to be proportional to W. We therefore write

(1.3) $$F = f(\theta) \cdot W.$$

The problem is to specify $f(\theta)$. See Figure 1.27.

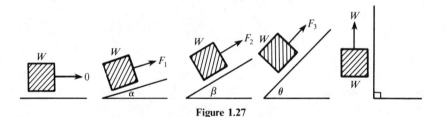

Figure 1.27

Let ρ be the density of the rope, i.e. the weight per unit length; then the weight of the rope AB is $a\rho$ and that of BC is $b\rho$. Thus, if AB is inclined at angle α to the horizontal, the force F_1 needed to prevent it slipping down is given by

(1.4) $$F_1 = a\rho \cdot f(\alpha).$$

Likewise, the force F_2 needed to prevent BC slipping down its incline of angle β is

(1.5) $$F_2 = b\rho \cdot f(\beta).$$

But since the rope AB counterbalances the rope BC

$$F_1 = F_2;$$

so by (1.3), (1.4), and (1.5),

$$a\rho \cdot f(\alpha) = b\rho \cdot f(\beta) \quad \text{or} \quad af(\alpha) = bf(\beta).$$

From the geometry of Figure 1.26,

$$a = \frac{h}{\sin \alpha}, \quad b = \frac{h}{\sin \beta},$$

giving

$$\frac{h}{\sin \alpha} \cdot f(\alpha) = \frac{h}{\sin \beta} \cdot f(\beta)$$

so that

$$\frac{f(\alpha)}{\sin \alpha} = \frac{f(\beta)}{\sin \beta}.$$

But Stevin's argument is applicable to any arbitrary triangle ABC. No matter what non-obtuse angle α we have selected for the one incline, we are free to select β for the other incline *independently* of our first choice. If we take another case of Fig. 1.26 with angles α, β' we similarly deduce

$$\frac{f(\alpha)}{\sin \alpha} = \frac{f(\beta')}{\sin \beta'}$$

giving, with our first result,

$$\frac{f(\alpha)}{\sin \alpha} = \frac{f(\beta)}{\sin \beta} = \frac{f(\beta')}{\sin (\beta')}.$$

In other words, for any non-obtuse angle θ, $f(\theta)/\sin \theta$ is constant, say $f(\theta)/\sin \theta = C$. Hence, $f(\theta) = C \sin \theta$ and (1.3) becomes

(1.6) $$F = W \cdot C \cdot \sin \theta.$$

It remains to determine C. When $\theta = 90°$, W is as if suspended adjacent to a vertical wall, and clearly $F = W$. Substituting in (1.6), we have

$$W = W \cdot C \cdot \sin 90° = W \cdot C \cdot 1,$$

therefore

$$C = 1 \quad \text{and} \quad F = W \sin \theta.$$

1.7 Insight and Outlook

In this chapter we have very briefly sketched the development of mechanics by first examining its origin with Archimedes' work on the lever and then considering its subsequent development by Galileo and Stevin. Yet it must not be supposed that we advocate an historical approach to the study of science. Our concern is not to retrace in detail the thousand and one errors of the past; past mistakes are of interest to us only insofar as they illuminate subsequent successes. On the contrary, our approach is genetic; primarily to retrace only the great strides forward. Broadly speaking, the work of first rate men of one era serves as a foundation for the work of first rate men of the succeeding era. The sequence in which fruitful concepts evolved is a first indication of the sequence in which to study them. To follow in the footsteps of giants is to wear seven-league boots.

Ponder in particular the inscription Stevin gave his diagram: "It looks like a miracle, but it is not a miracle." The endless rope which does not slide upon the triangle contains, so to speak, the law of the inclined plane.

By using the genetic approach, by going back to the beginnings of science, to Archimedes, to Galileo and to Stevin, we begin to get some idea of how able mathematicians go about their business. We have an indication of how simple ideas, relentlessly applied with high imagination, are developed into a theory. Here, theorems are exposed for what they are: conquests of intuition. To do mathematics, study mathematics in the making.

CHAPTER TWO

Growth Functions

Hats, therefore hat pegs; growth, therefore growth functions. What could be a more natural introduction to the concept of function than growth problems? In the first section we show how the exponential law of growth is derived from a functional equation that arises naturally from its context. In the second section we consider an application by Maxwell of this result. Next, in considering population growth, we are led from functional to recursive equations; solutions behave in a surprisingly complicated way—a subject of much current research. In Section 5 recursion equations describe the "growth" of the side length of a regular polygon of fixed perimeter, thus leading to Cusanus' formula for π. The chapter ends with a proof and applications of the arithmetic-geometric mean inequality.

2.1 The Exponential Law of Growth

How much timber is there in a forest? Trees grow. The older the forest, the bigger the trees. The bigger the trees, the greater the amount of wood. Provided that there are no forest fires and no trees die, the volume of wood increases with time. The volume of timber depends upon how long the forest has been growing; it is a function of the time for which the trees have been growing.

Doesn't this situation invite introduction of a mathematical notation? We introduce one. Let $N(t)$ denote the volume of timber (in cubic feet, say) in a given forest t years after it was planted. When $t = 0$, $N(t) = N(0)$; that is, $N(0)$ is the initial amount of timber in the forest when it was planted.

We take it to be an axiom that identical forests have identical amounts of wood; that, in other words, to double the initial acreage is to double the amount of wood, to treble it is to treble the amount of wood,.... While $N(0)$ grows to $N(t)$, a forest of initial size $2N(0)$ grows to $2N(t)$, a forest of initial size $3N(0)$ grows to $3N(t)$,.... Our axiom implies:

$$(2.1) \qquad \frac{N(t)}{N(0)} = \frac{2N(t)}{2N(0)} = \frac{3N(t)}{3N(0)} = \cdots.$$

Let g be the growth ratio (2.1) for a forest with an initial amount of timber $N(0)$ for a growing period of t years; i.e.

$$(2.2) \qquad \frac{N(t)}{N(0)} = g.$$

It immediately follows from (2.1) that for a fixed growing period t all forests have the same growth ratio, namely g, irrespective of their initial amount of timber. In short, (2.2) holds for all forests.

So far, so good. But the fact that g is the same—a constant—for all forests for *a given growth period* must not mislead us into supposing that g is the same—a constant—for *different growth periods*. On the contrary, for as long as any forest continues to grow, the longer its growing period, the greater its growth and therefore the greater its growth ratio (2.2). As t increases, g increases. Although for given t, g is the same for all $N(0)$, it is different for different t. To record the dependence of g on t we set $g = f(t)$ and write (2.2) in the form

$$(2.3) \qquad N(t) = N(0) \cdot f(t).$$

With hindsight, (2.3) is obvious. If $f(t)$ is the amount of wood in one tree after t years of growth, then $N(0)$ trees growing for the same period have a total amount of wood $N(0) \cdot f(t)$, the amount $N(t)$ in the whole forest.

Now that we have given a mathematical formulation based on common knowledge of uninhibited growth, we have a formula that we can sink our teeth into.

Suppose for ease of exposition that our forest was planted at the turn of the century. Then 5 years later, in 1905, its size was $N(5)$, where

$$(2.4) \qquad N(5) = N(0)f(5).$$

What is its content in 1911? What, in other words, is the content (5 + 6) years after planting? Two ways of answering this question now present themselves: one in terms of its growth since it was planted in 1900,

GROWTH FUNCTIONS

(5 + 6) years earlier; the other in terms of its additional growth since 1905, 6 years earlier. By (2.3) the first answer is

(2.5) $$N(5 + 6) = N(0)f(5 + 6).$$

The second answer is slightly less obvious. We consider our forest as if it were initially observed in 1905 with the initial size $N(0) \cdot f(5)$—see (2.4) —and had grown for only 6 years. By (2.3), we have

(2.6) $$\overline{N}(6) = [N(0) \cdot f(5)] f(6)$$

(the bar in "$\overline{N}(6)$" is used to remind us that "6" refers to 6 years after 1905, not 1900). But these two answers, given by (2.5) and (2.6), must be the same; for $N(5 + 6)$, the volume of wood in our forest $(5 + 6)$ years after 1900, is the volume $\overline{N}(6)$ of wood 6 years after 1905. Consequently,

$$N(0) \cdot f(5 + 6) = [N(0) \cdot f(5)] f(6),$$

which gives the *functional equation*

$$f(5 + 6) = f(5) \cdot f(6).$$

The specific periods, 5 and 6 years, were used for ease of exposition. The argument may be repeated with arbitrary periods t_1, t_2, giving the functional equation

(2.7) $$f(t_1 + t_2) = f(t_1) \cdot f(t_2).$$

Reflect that to deduce this equation we did not need any technical knowledge of biology or forestry, but merely to articulate what hitherto we had never stopped to think about. To get started was to stop to think.

Let us use the functional equation (2.7) to determine $f(t)$. Putting $t_1 = t_2 = t$, we find

$$f(t + t) = f(2t) = f(t) \cdot f(t) = [f(t)]^2,$$

and with $t_1 = t$, $t_2 = 2t$,

$$f(3t) = f(t) \cdot f(2t) = f(t) \cdot [f(t)]^2 = [f(t)]^3.$$

These results lead us to suppose that

(2.8) $$f(nt) = [f(t)]^n$$

for every positive integer n. This clearly holds for $n = 1$. If it holds for some positive integer n, then

$$f((n+1)t) = f(nt + t) = f(nt) \cdot f(t) = [f(t)]^n f(t) = [f(t)]^{n+1},$$

so that (2.8) holds also for $n + 1$. Consequently, by the principle of mathematical induction, (2.8) holds for every positive integer n.

Thus we have

(2.8) $$f(nt) = [f(t)]^n,$$

(2.9) $$f(mt) = [f(t)]^m,$$

where n, m are positive integers. Putting $t = 1/n$ in (2.8) yields

$$f(1) = [f(1/n)]^n;$$

taking the n-th root yields

$$[f(1)]^{1/n} = f(1/n),$$

and raising to the m-th power,

$$[f(1)]^{m/n} = [f(1/n)]^m.$$

Now we set $t = 1/n$ in (2.9) and find that

$$f(m/n) = [f(1/n)]^m.$$

Therefore,

$$f(m/n) = [f(1)]^{m/n}.$$

Putting $m/n = t$ and $f(1) = a$, we obtain the formula

(2.10) $$f(t) = a^t,$$

where a, the value of $f(1)$, is a constant, and t is any positive rational since m, n are arbitrary positive integers. This is known as an *exponential function*.

By (2.3) the law of growth for our forest becomes

(2.11) $$N(t) = N(0) \cdot a^t,$$

the value of a depending upon the kind of forest considered.

GROWTH FUNCTIONS

(Strictly speaking $f(t)$ has been defined only for rational values of t; but if it is conceded that a forest grows continually, then (2.11) holds for all (real) values of t. A rigorous discussion of this point can be found in calculus books.)

We have answered the question: How much timber is there in a forest? It takes but slight reflection to see that the law of growth need not be applicable solely to forests. Of course, it is applicable to any phenomenon where growth occurs like that of trees.

Our derivation of the law of exponential growth for forests depended on two properties of the size of forests as function of time:

(i) That the size, after time t, is proportional to its initial size; this property is embodied in equation (2.3).

(ii) That the rate of growth obeys the same law at all times; that is, the initial time 0 in equation (2.3) may be replaced by any other time s:

$$N(s + t) = N(s)f(t).$$

As we saw above, this relation leads to the functional equation (2.7) for f whose solutions are exponential functions (2.10).

A forest is a population of trees. The law of exponential growth applies to any population having properties (i) and (ii) above. There are many such, e.g., a population of bacteria; here property (ii) is satisfied as long as all conditions, such as the availability of nutrients, remain the same. Another example is a population suffering from a specific communicable disease. Under certain conditions, called *epidemic*, such a population has the two properties above and therefore grows exponentially. You may have read in the paper or heard in a news broadcast that the number of victims of AIDS doubles every six months.

Exercise 2.1. If the number of AIDS victims doubles every six months, and if there are 1500 sufferers today, how many will there be 5 years from now?

In Section 2.4 we shall discuss conditions of population growth under which property (i) is invalid and must be modified. This will lead to growth laws quite different from the exponential law.

So far we have spoken of exponential *growth*; but our discussion applies equally well to *decreasing* populations. An important example is radioactive decay. According to the theory, substantiated by experiment, in a given unit of time a certain fraction of radioactive atoms decay. Because this decay is a completely random process, among twice as many atoms twice as many decay; this leads to property (i). Here property (ii) asserts that the laws of physics governing radioactive decay are immutable. It follows therefore that radioactive decay occurs at an exponential rate given by formula (2.10), but now a is a positive number *less than* 1.

The exponential nature of radioactive decay gives rise to the concept of *half life*, the time T it takes for a given amount of radioactive material to shrink to half its size.

Exercise 2.2. (a) Show that if a radioactive substance is governed by formula (2.10), the half life T of that substance satisfies

$$a^T = 1/2.$$

(b) The half life of radium is 1,622 years; what is the value of a for radium?

The intensity of a ray of light passing through an absorbing medium decays exponentially *as function of the distance it traverses through the absorbing layer*. Here property (ii) expresses the fact that the optical properties of the medium are uniform. Thus we have

$$I(x) = I(0)a^x,$$

where $I(0)$ is the intensity of the incident light ray at the surface of the absorbing medium, $I(x)$ the intensity at a distance x from its point of entry into the absorbing medium, and a (less than 1) the absorption factor.

Growth of money invested at interest which is periodically added to the capital is another example. This is called *compound* interest, in contrast to *simple* interest investments where the capital remains fixed because the earned interest is *not* added to the initial investment. Compound interest investments grow exponentially. Here property (i) expresses the fact that the interest paid to depositors is proportional to the amount they have on deposit; property (ii) says that the interest rate remains the same.

Let p denote the annual interest rate (in percents). If interest is compounded annually, then after 1 year a capital C_0 will have grown to

$$C_1 = C_0 + \frac{p}{100}C_0 = C_0\left(1 + \frac{p}{100}\right) = C_0 a, \qquad a = 1 + \frac{p}{100}.$$

Since one year is like any other, after 2 years the capital C_0 will have grown to

$$C_2 = C_1\left(1 + \frac{p}{100}\right) = C_1 a = C_0 a^2,$$

GROWTH FUNCTIONS

and after n years it will have grown to

$$C_n = C_0 a^n.$$

What happens when interest is compounded every half year? After half a year the interest paid is only half that for a full year, so the capital C_0 after half a year will have grown to

$$C_0 + \frac{1}{2}\frac{p}{100}C_0 = C_0\left(1 + \frac{p}{200}\right);$$

after a full year, C_0 has grown to $C_0(1 + p/200)^2$.

Of course interest can be—and has been—compounded quarterly, monthly, weekly, daily. We shall examine the effect of increasing the frequency of compounding interest on the growth of capital.

For sake of simplicity we take the interest rate p to be 100 percent, a usurious one under normal circumstances (but not so far out in one of the wildly inflationary economies that plague, alas, many countries today). If compounded at the end of the year, the capital doubles, i.e. increases by a factor of 2. If compounded semiannually, at the end of the year the capital will have increased by a factor $(1 + \frac{1}{2})^2 = 2.25$. If compounded quarterly, it will have increased by a factor $(1 + \frac{1}{4})^4 \approx 2.4414$. Let us denote by A_N the factor by which the capital increases in one year if it is compounded at N equal intervals, at 100% annual interest; then

$$A_N = \left(1 + \frac{1}{N}\right)^N = \left(\frac{N+1}{N}\right)^N.$$

Here is a list of values of A_N for $N = 1, 2, \ldots, 12$, computed to three decimal places:

$A_1 = \left(\frac{2}{1}\right)^1 = 2.000$ $A_5 = \left(\frac{6}{5}\right)^5 \approx 2.448$ $A_9 = \left(\frac{10}{9}\right)^9 \approx 2.581$

$A_2 = \left(\frac{3}{2}\right)^2 = 2.250$ $A_6 = \left(\frac{7}{6}\right)^6 \approx 2.522$ $A_{10} = \left(\frac{11}{10}\right)^{10} \approx 2.594$

$A_3 = \left(\frac{4}{3}\right)^3 \approx 2.370$ $A_7 = \left(\frac{8}{7}\right)^7 \approx 2.546$ $A_{11} = \left(\frac{12}{11}\right)^{11} \approx 2.604$

$A_4 = \left(\frac{5}{4}\right)^4 \approx 2.441$ $A_8 = \left(\frac{9}{8}\right)^8 \approx 2.566$ $A_{12} = \left(\frac{13}{12}\right)^{12} \approx 2.613$

For curiosity, we add the values A_{52} and A_{365} corresponding to weekly and daily compounding:

$$A_{52} = \left(\tfrac{53}{52}\right)^{52} \approx 2.693, \qquad A_{365} = \left(\tfrac{366}{365}\right)^{365} \approx 2.715.$$

Now let us tabulate a few values of A_N for very large N:

N	10^3	10^4	10^5	10^6	10^7	10^8	10^9
A_N	2.7169239	2.7181459	2.7182682	2.7182804	2.7182817	2.71828181	2.71828182

These numbers strongly suggest that the sequence $A_1, A_2,\ldots$ increases and tends to some limiting value. By "tending to a limiting value", a concept of the calculus, we mean that all but a finite number of members of the sequence have the same first k digits, no matter what k is.

In Section 2.6 we shall pose exercises designed to prove the increasing nature and the convergence of the sequence A_N. Here we note that both are intuitively clear; for, the more frequently interest is added to our capital, the more money we have at the end of the year. On the other hand, we cannot expect to become infinitely rich, no matter how frequently the interest is compounded; so increasing the frequency of compounding cannot make much difference beyond a certain large enough frequency.

The limit of $A_1, A_2, \ldots$ is denoted by e:

$$(2.12) \qquad \lim_{N \to \infty} A_N = \lim_{N \to \infty} \left(1 + \frac{1}{N}\right)^N = e \approx 2.71828\ldots$$

It is one of the most important numbers in mathematics, introduced by Napier in the 17-th Century as base of natural logarithms.

As our tables show, the sequence $\{A_N\}$ converges to e rather slowly. We next sketch a much more efficient way of computing e to any desired accuracy.

We recall the binomial theorem:

$$(a+b)^n = \sum_{k=0}^{n} \binom{n}{k} a^{n-k} b^k,$$

where n and k are non-negative integers, $k \leq n$, and the binomial coefficients $\binom{n}{k}$ are defined by

$$\binom{n}{k} = \frac{n(n-1)(n-2) \cdots (n-(k-1))}{k!}.$$

Now let $a = 1$, $b = 1/n$, so that

$$\left(1 + \frac{1}{n}\right)^n = \sum_{k=0}^{n} \binom{n}{k} \frac{1}{n^k}$$

$$= 1 + \frac{n}{1} \cdot \frac{1}{n} + \frac{n(n-1)}{2!} \cdot \frac{1}{n^2} + \cdots + \binom{n}{k} \frac{1}{n^k} + \cdots + \frac{1}{n^k}.$$

We write the $(k+1)$-th term in the form

$$\binom{n}{k}\frac{1}{n^k} = \frac{1}{k!}\frac{n-0}{n}\cdot\frac{n-1}{n}\cdot\frac{n-2}{n}\cdot\ldots\cdot\frac{n-(k-1)}{n}$$

$$= \frac{1}{k!}\cdot 1\left(1-\frac{1}{n}\right)\left(1-\frac{2}{n}\right)\cdots\left(1-\frac{k-1}{n}\right),$$

As n increases, the numbers $1-(s/n)$ approach 1; hence

$$\lim_{n\to\infty} A_n = \lim_{n\to\infty}\left(1+\frac{1}{n}\right)^n = \lim_{n\to\infty}\frac{1}{n^k}\binom{n}{k} = \sum_{k=0}^{\infty}\frac{1}{k!} = e;$$

thus

$$e = \frac{1}{0!}+\frac{1}{1!}+\frac{1}{2!}+\cdots = 1+1+\frac{1}{2!}+\frac{1}{3!}+\cdots.$$

We now return to arbitrary interest rates p and arbitrary times t of deposits. When compounded at N equal intervals during the time period t, the original capital will increase by the factor

$$\left(1+\frac{p}{100}\frac{t}{N}\right)^N.$$

Set $M = \dfrac{100N}{pt}$, and rewrite this factor as

$$\left(1+\frac{1}{M}\right)^{M(pt/100)} = \left[\left(1+\frac{1}{M}\right)^M\right]^{pt/100}.$$

Since $(1+1/M)^M$ tends to e, see (2.12), it follows that the above factor tends to

$$e^{pt/100}$$

as M increases without bound. This is the factor by which our capital increases if it earns interest at p percent per year for t years and is, so to speak, *compounded instantaneously*, or *continuously*.

The concept of continuous compounding, natural to a mathematician[†], used to be alien to bankers. They were led to consider it in their fierce

---

[†] The first to consider continuous compounding was the famous Swiss mathematician James Bernoulli (1690).

competition to compound interest on deposits more frequently than their competitors.

To recapitulate: We have shown how the concept of function and the barest rudiments of functional equation theory may be used to deduce the exponential law of growth, and we have indicated some fields of application.

2.2 Maxwell's Derivation of the Law of Errors

In this section we consider Gauss' law of errors (Gauss 1777–1855). We shall find that Maxwell's (Maxwell 1831–1879) ingenious derivation of it depends upon the solution of a functional equation. This solution is an application of the functional equation for the exponential function considered in the preceding section.

When at the beginning of the last century astronomers, physicists and surveyors started to make precise measurements, it was realized that there is no such thing as an absolutely accurate measurement.

First consider the question of a single observation. Astronomers chart the stars as accurately as they know how, yet two astronomers seldom observe the same star as being in the same position. The figures expressing their measurements are apt to differ in the last decimal place or two.

To come nearer home; the spring in your bathroom scale becomes fatigued and loses a little of its springiness. With changes in temperature, bits of metal alter in length and so modify the mechanism. If over-conscientious about your weight, you may evade many of these contributions to inaccuracy by resorting to an equal-arm balance of appropriate dimensions. But even the arms of balances become tired and droop a little. Better designed and more carefully constructed instruments measure more accurately, but there are no absolutely accurate measuring devices. We include the human eye reading a pointer against a graduated scale.

We suppose you, afflicted by a weight-reducing fad, weighed yourself on three bathroom scales this morning, their readings being 201, 207, 204 pounds. Your problem: What was my weight this morning? Possibly you would, in the absence of a known weight with which to test the scales, take the arithmetic average, 204, pounds, as correct. You would conclude almost with certainty that you did not weigh 300 pounds, and think it very unlikely that you were as much as 250. Surely the further removed the estimated figure from 204 or thereabouts, the more unlikely its correctness.

Uninvited, the notion of probability intrudes upon the scene. Uncertain of the correct figure we cannot be certain of the error of the measured reading; the most we can ask is such questions as, "What is the likelihood that the observed reading n does not differ from the actual measure by, say, more than $.01n$?" The general answer to questions of this sort is called the law of errors. With this answer we shall be presently concerned.

Secondly, consider the question of the combination of observations. Although hundreds of physicists have made measurements from which to deduce the velocity of light, no two physicists have obtained exactly the same result. The deduced number being dependent upon several measurements, each subject to error, the final result necessarily incorporates a combination of these errors.

Consider, for simplicity, the following example. A square lamina of side 5 units is measured as being a rectangle having adjacent sides of 5.1 and 4.9 units. So whereas the actual area is 25 square units, the area deduced on the basis of our measurements is 24.99. Although there is a 2% error in each of our measurements, there is only a $1/25$ of 1% error (.04%) in the final result. One measurement was too big, the other too small, so that each error tends to annul the inaccuracy due to the other. But this oversimplifies; the point is that we never know with certainty the actual errors. A more realistic question is: If it is 95% certain that the error in each of our measurements does not exceed 2%, what is the probability that the error in the area calculated on the basis of these measurements does not exceed, say, 1%?

We have briefly indicated the kind of problem this line of thought leads to; now we must return to what it leads from, the probability of such and such an error in a single measurement. As we have said, the general answer to this latter question is known as the law of errors.

The law was first derived by Gauss in the masterly way characteristic of this great mathematician; but his approach to the problem was so abstract that Maxwell, among others, was only partially convinced of the correctness of his derivation. It lacked that down-to-earthness found in, for example, Stevin's deduction of the law of the inclined plane. Maxwell was led to examine Gauss' proof when he needed the law of errors to further develop statistically the kinetic theory of gases. He was concerned with the down-to-earth conception of the behavior of a gas as that of billions of molecules darting to and fro, pushing against the walls of their enclosure, so it is perhaps not too surprising that he came up with a marvelous, immediately graspable proof. Yet on second thought it is most surprising, for although many contemporary physicists shared with

Maxwell his dissatisfaction with Gauss' formulation, none shared with him his discovery. Such is the mark of genius.

From the problem of molecules impinging on the walls of their enclosure, Maxwell turned to that of bullets hitting a target. Let us consider his derivation of the law of errors.

Consider the marksman who misses the bull's-eye. Typically, the (printable) phrase he uses to describe his shot is one of the following sort: to the right of center; left of center; above center; to the right of center and too high; left of center and low. He refers to his bullet's position as a combination of two errors; a horizontal and a vertical deviation from the bull's-eye. Taking our cue from him, we introduce rectangular coordinate axes with origin at the bull's-eye and x-axis horizontal. $H(x, y)$ is the position of his hit.

If a marksman is standing in a fixed position at a certain distance from the target, what is his probability of hitting the bull's-eye? First, this will depend upon the size of the bull's-eye. Surely we are agreed that if it is no bigger than the point of a pin, then it is practically impossible to hit; and that if it is conceived of as a mathematical point, then the probability of hitting it is zero. Thus we must reformulate our question: instead of asking, "What is the probability of hitting $(0, 0)$?" we must ask, "What is the probability of hitting a given neighborhood of $(0, 0)$?" The general question is, "What is the probability when aiming at (x, y) of hitting a certain neighborhood of (x, y)?"

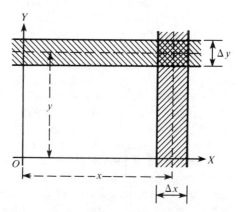

Figure 2.1

Obviously, the probability will depend upon the size of the neighborhood; take the whole world for the neighborhood of $(0, 0)$, and the marksman cannot miss. The neighborhood must be specified. It is natural

GROWTH FUNCTIONS

to take a rectangle of sides Δx, Δy, centered on (x, y) as the neighborhood of (x, y). See Figure 2.1. Yet there remains a question. What, explicitly, do we mean by "probability"? If a marksman in firing his first 1000 rounds at the bull's-eye hits a certain neighborhood 60 times, does the same thing with his second 1000 shots, with his third, and fourth thousand, then we would say that his probability of hitting that neighborhood is 60/1000. But it is commonplace that performance varies, even for an enthusiast whose marksmanship does not improve with practice. It would be more realistic to suppose his successive scores $60, 57, 62, 59, \ldots$. To judge his expectation of hitting the neighborhood we would consider his performance in the long run. More generally, the probability of a shot hitting a neighborhood of (x, y) will be said to be p, if this neighborhood is hit pn times with n shots, where n is very large.

Clearer as to what we mean by "probability," we readdress the question: "What is the probability of a hit in the rectangular $\Delta x \times \Delta y$ neighborhood of (x, y)?" For brevity, we denote this probability by $P(x, y, \Delta x, \Delta y)$. But aren't we really asking two questions? Or, to be more precise, are there not two (easier) questions on which the answer to our original question depends? (1) What is the probability that a hit will lie in the vertical strip of width Δx centered on x? Symbolically, $P(x, \Delta x)$? (2) What is the probability that a hit will lie in the horizontal strip of width Δy centered on y? Symbolically, $P(y, \Delta y)$? Study the conjunction of Figures 2.2a, 2.2b to give Figure 2.1. Does not this make it clear that our original question may be construed as: What is the probability that a hit will lie in *both* strips?

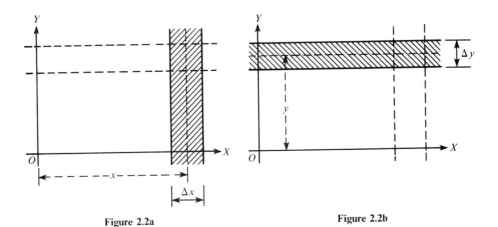

Figure 2.2a Figure 2.2b

How, specifically, does $P(x, y, \Delta x, \Delta y)$ depend on $P(x, \Delta x)$ and $P(y, \Delta y)$? The dependence may be illustrated by a problem of throwing dice.

Suppose that the probability of throwing a 3 with a given die is $\frac{1}{6}$ and that of throwing a 4 with a second die is also $\frac{1}{6}$. What is the probability of throwing a 3 with the first and a 4 with the second? In the long run 3 turns up with a frequency of $\frac{1}{6}$, so that if we consider $36n$ throws of two dice, where n is large, $6n$ of these throws will have a 3 uppermost on the first die (and the other $30n$ throws will not). Of these $6n$ throws, since the frequency of a 4 with the second die is $\frac{1}{6}$ of the number of throws, independently of what comes up on the first die, just n of them will have a 4 uppermost. Thus, just n of the $36n$ pairs will have a 3 uppermost on the first and a 4 uppermost on the second. In short if two independent events have probabilities of $\frac{1}{6}$ and $\frac{1}{6}$, then the combined event has a probability of $\frac{1}{6} \times \frac{1}{6}$. More generally, if p_1, p_2 are the probabilities of two independent events, then the probability of the combined event is $p_1 \times p_2$, the product of the individual probabilities. It follows that

$$(2.13) \qquad P(x, y, \Delta x, \Delta y) = P(x, \Delta x) P(y, \Delta y).$$

It is an open mathematical secret that with two questions to answer it is best to answer them one at a time. What is $P(x, \Delta x)$? If a distant barn is five times as wide as its door, then it seems reasonable to suppose that the chance of hitting the barn is five times that of hitting the door. Mutatis mutandis, if the door is fixed in position (i.e., the position x of its center line is fixed, say $x = x_1$) but its width Δx varies, then the probability of hitting it from a long distance away varies directly as its width. So, we take it as an axiom that for a vertical strip whose center line is $x = x_1$,

$$(2.14) \qquad P(x_1, \Delta x) = k_1 \Delta x$$

where k_1 is independent of Δx.

But, although the "constant of proportionality" is independent of the width Δx of the vertical strip, it is obviously not independent of the position of the strip (i.e., the abscissa x_1 of its center line). Consider, for example, a barn with two doors of the same size. Surely the chances of hitting the one we aim at, straight in front of us, is greater than that of hitting the other. The farther to the side the other is, the smaller its chance of being hit. Thus, reminiscent of (2.1) and (2.2) we will have

$$P(x_1, \Delta x) = k_1 \Delta x$$
$$P(x_2, \Delta x) = k_2 \Delta x$$

GROWTH FUNCTIONS

exemplifying the pattern

$$P(x_n, \Delta x) = k_n \Delta x,$$

where k_n, although unchanged by changes in Δx, is dependent upon, i.e., is a function of, x_n. In short,

(2.15) $$P(x, \Delta x) = F(x) \cdot \Delta x.$$

Suppose a barn to have three doors of the same size, the one to the left and the one to the right being equidistant from the one straight ahead of us. Surely the chance (when aiming at the middle one) of hitting the one on the left is the same as that of hitting the one on the right. The chances of a "left" error are the same as those of an equal "right" error. Mathematically,

$$P(x, \Delta x) = P(-x, \Delta x).$$

Hence, by (2.15)

$$F(x) = F(-x),$$

that is, $F(x)$ is an *even function*.

Since $(x)^2 = (-x)^2$, clearly any function $f(x^2)$ of x^2 is an even function of x. There is, for example, no gain in generality in taking $f(x^4)$, or $f(x^6)$, for these are also of the form $f(X^2)$ with $X = x^2$, $X = x^3$, respectively. Thus (2.15) takes the form

(2.16) $$P(x, \Delta x) = f(x^2) \cdot \Delta x$$

which indicates, for example, that the probability of a hit in the left-hand strip of Figure 2.3 is the same as that of a hit in the right-hand strip.

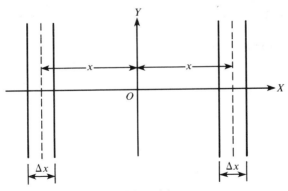

Figure 2.3

The next question is: What is $P(y, \Delta y)$? Again compare Figure 2.2a with Figure 2.2b. What differences are there? If $x = y$ and $\Delta x = \Delta y$, the strips are of the same size and at the same distance from O. The only difference is that of direction; the one is above, the other to the right of, O. And what role does difference of direction play? We are agreed that a hit, (say) 3 inches left of center has the same probability as a hit 3 inches right of center; is a hit 3 inches above center more likely than 3 inches below center? Right of center was given no preference over left of center; why should above center be given preference over below center? It is natural to consider them equiprobable. This leads to another question: Is a hit 3 inches to right of center more likely than, say, 3 inches above center? Consider the circle of radius 3 inches with center O, illustrated by Figure 2.4.

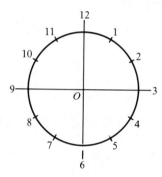

Figure 2.4

In firing at O, (the immediate neighborhood of) which point on this circle has the greatest probability of being hit? Has the point at "1 o'clock" more or less probability of being hit than that at "2 o'clock"? We suppose the probability of hits at any two points on this circle to be equal; no direction is preferred.

Direction being considered irrelevant, it follows that the strips of Figure 2.2a, 2.2b (with $x = y$, $\Delta x = \Delta y$) are not only of the same width and at the same distance from O, but they are also similarly situated with respect to O in the probabilistic sense. Thus $P(y, \Delta y)$ is determined by precisely the same function as $P(x, \Delta x)$. So, by (2.16)

(2.17) $$P(y, \Delta y) = f(y^2)\Delta y$$

and hence by (2.13)

(2.18) $$P(x, y, \Delta x, \Delta y) = f(x^2)f(y^2)\Delta x \Delta y.$$

GROWTH FUNCTIONS

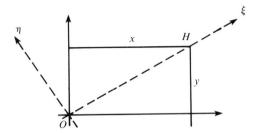

Figure 2.5

At this stage Maxwell displays his ingenuity. He rotates the x- and y-axes about the origin; the x-axis goes into the ξ-axis, and the y-axis into the η-axis of a new coordinate system. See Figure 2.5. His ingenuity is displayed in making the ξ-axis go through the point H whose old coordinates are (x, y); the ordinate of H relative to the new axes is zero, for H lies on the ξ-axis. $H(x, y)$ relative to the old axes becomes $H(\xi, 0)$ relative to the new. And since the probability of a hit within the neighborhood of a point is independent of the direction of the axes to which it is referred, the probability of a hit within a neighborhood of H is given by

(2.19) $$P(\xi, 0, \Delta\xi, \Delta\eta) = f(\xi^2)f(0)\Delta\xi\,\Delta\eta$$

as well as by (2.18). From (2.18), (2.19), we have,

$$f(x^2)f(y^2)\Delta x\,\Delta y = f(\xi^2)f(0)\Delta\xi\,\Delta\eta.$$

Both members of this equation give the probability of hitting an area element $A = \Delta x\,\Delta y$ or $A = \Delta\xi\,\Delta\eta$. Since the probability density is $f(x^2)f(y^2)$ in the first case and $f(\xi^2)f(0)$ in the second, we may conclude that

$$f(x^2)f(y^2) = f(0)f(\xi^2).$$

By Pythagoras' Theorem, $\xi^2 = x^2 + y^2$, so

(2.20) $$f(x^2)f(y^2) = f(0)f(x^2 + y^2).$$

Equation (2.20) is a functional equation of the form

(2.21) $$f(a)f(b) = Kf(a + b),$$

where $f(0) = K$. Here our imagination may be sparked, for we recall that the functional equation for the law of growth is of the form

(2.22) $$f(a)f(b) = f(a+b).$$

If K were equal to 1, then the law of errors would have the same functional equation as the law of growth, and consequently the solution of (2.20) would likewise be an exponential function.

We can reduce (2.20) to the form (2.22) by setting

$$g(x^2) = \frac{f(x^2)}{f(0)}$$

so that

$$f(x^2) = f(0)g(x^2),$$
$$f(y^2) = f(0)g(y^2),$$
$$f(x^2 + y^2) = f(0)g(x^2 + y^2).$$

Substituting these expressions into (2.20), we have

$$[f(0)g(x^2)][f(0)g(y^2)] = f(0)[f(0)g(x^2 + y^2)].$$

Dividing by $[f(0)]^2$, we obtain

$$g(x^2)g(y^2) = g(x^2 + y^2),$$

the functional equation of the law of growth. Consequently

$$g(x^2) = a^{x^2} \quad \text{i.e.,} \quad \frac{f(x^2)}{f(0)} = a^{x^2}$$

so that

$$f(x^2) = f(0)a^{x^2};$$

and, by (2.16)

$$P(x, \Delta x) = f(0)a^{x^2}\Delta x.$$

For brevity, we set $f(0) = A$, and finally obtain

(2.23) $$P(x, \Delta x) = Aa^{x^2}\Delta x.$$

This completes Maxwell's derivation of Gauss' famous *law of errors*.

GROWTH FUNCTIONS

We discuss this law briefly. Since the chance of a large deflection is obviously smaller than the chance of a small deflection, we must have $a < 1$. Plotting a^{x^2} as a function of x, we obtain the bell-shaped curve associated with symmetrically deviating errors. See Figure 2.6. This is the starting point for the development of the whole theory of error in combinations of observations.

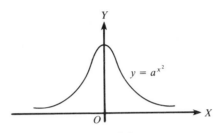

Figure 2.6

Gauss made an important application of this result. He was involved in the very practical problem of surveying his state of Hannover, and was disturbed by the fact that every measurement of a distance or an angle was subject to error. When a quantity x is measured n times, one generally obtains n different readings, $x_1, x_2, \ldots, x_n$, for the actual unknown value x. Each reading x_k means that the measurement falls in a small interval of length Δ centered at x_k, where Δ depends on the sensitivity of the measuring instrument. How does one make the best use of these measurements? Gauss considered the operation as a target practice where one tries to hit the target x, but rarely succeeds perfectly. In deriving equation (2.23), we considered the probability of hitting a neighborhood of a point x_k when we aimed at the bull's eye, O. With the bull's eye at x rather than at O, we must replace x_k by $x_k - x$ throughout the foregoing discussion. With this replacement in equation (2.23), we find that the probability of getting the reading for x_k is $Aa^{(x-x_k)^2}\Delta$. Since we assume that our measurements are independent of each other, the probability of the combined event of getting the n readings $x_1, \ldots, x_n$ is the product of the individual probabilities, which, by the laws of exponents, is

$$Ca^{(x-x_1)^2+(x-x_2)^2+\cdots+(x-x_n)^2},$$

where C is a constant. For what value of x is this event as probable as possible? Since $a < 1$, we should pick x so that the exponent is as small as possible. This leads to the celebrated *method of least squares*:

Given n measurements $x_1, x_2, \ldots, x_n$ of a quantity x, pick x so as to minimize

$$(2.24) \quad (x - x_1)^2 + (x - x_2)^2 + \cdots + (x - x_n)^2 = q(x).$$

When we learned how to "complete the square", we found that a quadratic function

$$q(x) = ax^2 + bx + c, \quad a > 0,$$

can be re-written

$$q(x) = a\left[\left(x + \frac{b}{2a}\right)^2 + \frac{c}{a} - \left(\frac{b}{2a}\right)^2\right],$$

so that, since $(x + b/2a)^2$ cannot be negative, $q(x)$ is minimal when $x = -b/2a$. For the quadratic function $q(x)$ of (2.24), $a = n$, and $b = -2(x_1 + x_2 + \cdots + x_n)$, so that $q(x)$ is minimal when

$$x = \frac{1}{n}[x_1 + x_2 + \cdots + x_n].$$

Alternatively we can use elementary calculus to find the minimizing x. Indeed,

$$q'(x) = 2[x - x_1 + x - x_2 + \cdots + x - x_n]$$
$$= 2n\left[x - \frac{1}{n}(x_1 + x_2 + \cdots + x_n)\right];$$

the minimum of $q(x)$ occurs when $q'(x) = 0$, hence is achieved for $x = 1/n(x_1 + x_2 + \cdots + x_n)$, the arithmetic mean of all n measurements. We conclude that the best value for x is the average of all the measurements obtained.

2.3 Differential and / or Functional Equations

Ever since Newton invented the calculus for his celestial mechanics, differential equations have been a principal tool of science. Hosts of examples could be cited, especially in physics, chemistry and more recently biochemistry to show how the statement of a natural law gives a differential equation whose solution furnishes information concerning natural phenomena. Thus it will come as no surprise to learn that examples of differential equations arising in mechanics will be found later in this book. Indeed, so fundamental is the calculus to science that to avoid giving such examples would be an unpardonable omission even though the book is addressed to the general reader. But in the next section, which deals with population growth, in order to get by with mathematics as elementary as possible, we shall not use differential equations even though they are appropriate to the investigation. Instead, as in the preceding sections of this chapter, we shall employ functional equations.

Whereas differential equations are generally easy to set up, functional equations are often hard to come by. And when they are available their solution often requires great ingenuity. Fortunately, the one we shall next encounter is "reasonable"; it can be dealt with in an elementary way. Later we shall see what can be done using both differential and functional equations, two strings to our bow.

2.4 The Problem of Predicting Population Growth

What is the law of increase of human population? The simplest, plausible assumption is that the population x_{n+1} of the $(n + 1)$th generation will be directly proportional to the population x_n of the nth. Symbolically,

$$(2.25) \qquad x_{n+1} = qx_n.$$

On this assumption, if x_1 is the population of the first generation considered, the population of successive generations will be $x_1, qx_1, q^2x_1, q^3x_1, \ldots$, so that

$$(2.26) \qquad x_n = q^{n-1}x_1.$$

Again we have an exponential law.

If $q > 1$ the population increases. Its rate of growth depends on the constant growth factor q; its value is influenced by characteristics of the human species and the environment as well as by social phenomena.

This formula was stated in words by Malthus (1766–1834):[†] populations of countries increase in geometric progression. It is interesting to

---

[†]Thomas Robert Malthus, *An Essay of the Principle of Population*, 1798, rev. ed. 1803.

note that Malthus was led to his formulation by inspecting the census records of the United States, which showed a doubling of population every 50 years. His statement, simple and crude as it was, had a tremendous influence on the whole of social philosophy in the 19th Century.

The social philosophers of the Enlightenment argued that it was mankind's duty to ease the hardship of the poor, and to abolish pestilence, plague, famine, and war, so that everyone could live happily till death of old age. Malthus thought this view greatly mistaken. What would happen with neither pestilence nor plague, with neither famine nor war? The population, increasing in geometric progression, would in the fullness of time, he argued, become so vast that the earth could not feed it; for the means of subsistence were found to increase in arithmetic progression. British industrialists in Manchester used this argument to prop up their policy of free enterprise, to increase trade while leaving the world at large to sort itself out. There could be no obligation to better the lot of the poor nor attempt to prevent famine or war; for these things, if evils, were evils necessary to prevent overpopulation. Malthus' law became the arithmetic of human misery.

Darwin also thought about the consequences of a population increasing geometrically. For him the problem had a wider context. He was at least as much interested in the increase of a colony of sea birds as in the Manchester birth rate. What, he asked, controls population? The dinosaur has long been extinct; the whale has survived. Ultimately, he gave an answer, his theory of natural selection. There followed his theory of evolution of species.

What is man's obligation to man? Is one to succor or to starve one's neighbor? By the middle of the last century even some industrialists began to question whether the existence of overworked and underpaid factory hands, living underfed in overcrowded slums, was a necessary evil. Couldn't there be a better arithmetic?

The Belgian sociologist, Verhulst (1804–1849), made an important observation. Catastrophes, wars, and plagues have occurred from time to time, but not all the time. Between any two successive catastrophes there was a period of tranquility, typically that of two or three generations.[†] This period, had the law of increase been geometric, would have given the population ample time, before the next catastrophe, to regain and surpass its size before the last, as illustrated in Figure 2.7. But mankind has inhabited the earth for thousands of years, so that although we do

---

[†]We may arbitrarily define the period of one generation, say, as 25 years and count the number of individuals on a fixed date every 25 years.

GROWTH FUNCTIONS

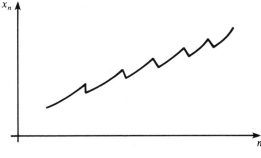

Figure 2.7

not know the value of n, where x_n is the present (nth generation) population, we do know that n is large. With n large, and the population before each catastrophe greater than that before the previous one, surely the world would now be overcrowded. Verhulst concluded that the geometric law does not give a correct account of the facts.

Discontent with the old arithmetic was the first step towards the new. Verhulst observed that human activities are of two kinds, cooperative and competitive. A marriage is the outcome of successful competition by a man against other men or a woman against other women for a mate; a child is the outcome of successful cooperation by a man and a woman. Farmers and biochemists cooperate to produce greater yields of wheat; bankers and bank robbers compete for the customers' deposits. Soldiers cooperate as armies to compete against other soldiers cooperating as armies. Verhulst took the view that in the main cooperation tends to increase, and competition tends to decrease, the population.

What is the effect of competition and cooperation on population growth? Verhulst considered the more frequent case where competition outweighs cooperation, and where encounters are more harmful than beneficial. In such situations population growth is stemmed by an amount proportional to the number of harmful encounters. Now the probability that any particular individual has a chance encounter with another is proportional to the number of people in the population. So the total number of encounters between all members of the population is, roughly, proportional to the square of the population size. Thus Verhulst arrived at a formula which today is known as *Verhulst's law*:

$$(2.27) \qquad x_{n+1} = qx_n - rx_n^2.$$

People who study changes in population size, called demographers, are eager to construct realistic models which not only describe the changes in population size over a known period, but also predict such changes for a future period. They are extremely eager to understand just what determines population size, and which of these influences might be controlled by social policies. In a model such as (2.27), how are the parameters q and r to be chosen and how do they depend on various characteristics of the population and its environment? How long should the time interval be between measurements x_n and x_{n+1} of a population size?

These are difficult questions. A model is successful if it predicts correctly the behavior that is observed. This has been the case for Verhulst's law applied to a wide class of populations. It is relatively easy for a mathematician to determine for what values of the parameters the model makes sense—and we shall do that for (2.27)—but it is a difficult matter to interpret, let alone control, these parameters.

By applying elementary mathematics to Verhulst's law (2.27) we shall extract information which matches observed phenomena about the ebb and flow of populations impressively well.

We simplify our work by eliminating the parameter r from (2.27) as follows: We multiply (2.27) by r, getting

$$rx_{n+1} = rqx_n - r^2x_n^2 = q(rx_n) - (rx_n)^2;$$

for rx_i we substitute y_i. Then in terms of the new quantities $y_i = rx_i$, Verhulst's law reads

(2.28) $$y_{n+1} = qy_n - y_n^2$$

and contains only one parameter, q. Whatever conclusions we reach concerning the sequence $y_1, y_2, \ldots$ can then be interpreted for the sequence $y_1/r = x_1, y_2/r = x_2, \ldots$. We can think of the y_i as just a re-scaling of the x_i; for brevity we refer to the y_i as population sizes.

We must restrict the parameter q to values for which the model is meaningful. Verhulst's law is meaningless if it predicts a negative population size. According to (2.28) y_{n+1} is negative only if $y_n > q$. Clearly the initial population size must not exceed q; but what is to prevent some future y_n from surpassing this barrier? It is Verhulst's law itself that keeps the population size bounded. To locate this bound we re-write (2.28) in the form

(2.28′) $$y_{n+1} = qy_n - y_n^2 = \frac{q^2}{4} - \left(y_n - \frac{q}{2}\right)^2.$$

The expression on the right never exceeds $q^2/4$, so that

$$y_{n+1} < q^2/4 \qquad \text{for all } n.$$

To make sure that no y_n exceeds q, it now suffices to require that

$$q^2/4 < q, \quad \text{i.e. that} \quad q < 4.$$

So from now on we assume that $q < 4$.

The $y_2, y_3, \ldots$ generated by (2.28) are positive and bounded by 4 provided that $0 < y_1 < q$ and $q < 4$. What else can we deduce about this sequence of numbers?

To answer this question we write (2.28) in the factored form

(2.28″) $$y_{n+1} = (q - y_n)y_n.$$

This resembles Malthus' law (2.25) except that the growth factor is now no longer a constant q; instead it is a function $G_n = q - y_n$ of the population size. It is always less than q. If q were less than 1, G_n would be still less and by (2.28″), $y_{n+1} = G_n y_n < y_n$ for all n. The population would dwindle to 0. To make growth possible we add the assumption

$$q > 1.$$

[Recall that even in the Malthusian case of growth unhampered by competition, there is growth only if $q > 1$.]

Since the growth factor $G = q - y$ plays such an important role in our analysis, we sketch it in Figure 2.8.

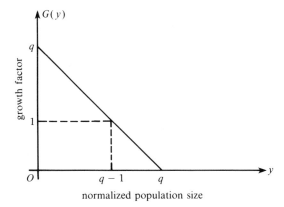

Figure 2.8

We identify the critical value of y for which the growth factor is exactly 1: $G(y) = q - y = 1$ when $y = q - 1$. This value is called *steady state* because for this number (2.28″) yields $y_{n+1} = y_n$, i.e. the population remains steady. We denote this critical size by s:

$$s = q - 1, \qquad G(s) = q - s = 1.$$

For $y_n < s$, $G_n > 1$ and $y_{n+1} = G_n y_n > y_n$; the population increases.

For $y_n = s$, $G_n = 1$ and $y_{n+1} = y_n$; the population remains fixed.

For $y_n > s$, $G_n < 1$ and $y_{n+1} = G_n y_n < y_n$; the population decreases.

Suppose the initial population size is below the steady state: $y_1 < q - 1$; will the sequence $y_1, y_2, \ldots$ increase beyond the steady state? Or suppose $q - 1 < y_1 < q$; will successive population sizes decrease beyond the steady state? Will each population be closer to s than the preceding?

Answers to these questions become visible when we write (2.28) in yet another form:

(2.28‴) $\qquad y_{n+1} = y_n + y_n[q - 1 - y_n] = y_n + y_n[s - y_n].$

Then

$$s - y_{n+1} = s - y_n - y_n[s - y_n] = (s - y_n)(1 - y_n),$$

and

(2.29) $\qquad\qquad \dfrac{s - y_{n+1}}{s - y_n} = 1 - y_n.$

The fraction on the left expresses the ratio of the deviation from s of the $(n + 1)$th population size to that of the nth.

If the value of this ratio lies beteen 0 and 1, that is if $0 < 1 - y_n < 1$, then by (2.29)

(i) successive populations cannot overshoot s because $s - y_{n+1}$ and $s - y_n$ have the same sign (their ratio > 0),
(ii) each y_{n+1} is closer to s than the preceding because the ratio of the absolute values of their differences from s is < 1.

Now if $0 < y_n < 1$, then $0 < 1 - y_n < 1$ and (i), (ii) would obtain.

Under what conditions will $y_n < 1$ hold for all n? We recall that $y_n \leq q^2/4$ for all n; so if we restrict the parameter q to

$$1 < q < 2,$$

GROWTH FUNCTIONS

then the population size cannot exceed

$$\frac{q^2}{4} < \frac{2^2}{4} = 1.$$

Consequently, for $1 < q < 2$ and any initial population size $y_1 < q$, we have $0 < y_n < 1$ for $n = 2, 3, \ldots$. Note also that in this case

$$0 < 1 - y_n < 1 \quad \text{and} \quad s = q - 1 < 1.$$

In summary: when $1 < q < 2$, an initial population below size s stays below s and grows toward s; an initial population above size s (but < 1) stays above size s and decreases toward s. Figure 2.9 shows these two possible behaviors, the only ones for $1 < q < 2$, except the steady case $s = y_1 = y_2 = \cdots$.

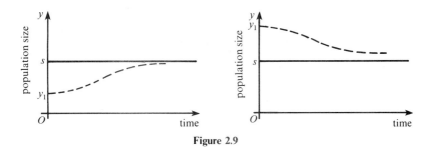

Figure 2.9

When, regardless of its initial size, a population tends to a steady state, that steady state is called *stable*. For a vast class of populations q is indeed less than 2 and the populations do tend toward the predicted steady state.

What does Verhulst's law predict for $q > 2$? Certainly, if $y_n = s$, all subsequent population sizes $y_{n+1}, y_{n+2}, \ldots$ are equal to s. What happens when y_n is close to, but not equal to, s? The answer can be read off (2.29): for y_n near s, $1 - y_n$ is near $1 - s$. Now $1 - s = 1 - (q - 1) = 2 - q$ which, for $2 < q < 3$ lies between -1 and 0. Therefore we deduce from (2.29) that for $2 < q < 3$,

(i) $s - y_{n+1}$ and $s - y_n$ have opposite signs,
(ii) $|s - y_{n+1}| < |s - y_n|$.

It follows that $y_n, y_{n+1}, \ldots$ tends to s in an oscillatory manner, being greater than s for, say, n even and less than s for n odd.

It is not hard to show that, as long as $2 < q < 3$, even if y_1 is far away from s, the sequence $y_1, y_2, \ldots$ eventually approaches s.

We have shown, with the help of (2.29), that for $2<q<3$, if y_n is near s, y_{n+1} is nearer still. The same argument shows that when $q > 3$, y_{n+1} is farther away from s than y_n. From this it follows easily that the sequence $\{y_n\}$ does *not* converge to the steady state in this case (unless it is exactly in the steady state for some n where it would then have to remain). In this situation the steady state is called *unstable*.

How does the sequence $\{y_n\}$ behave when q exceeds 3? Table 2.1 lists values of y_i obtained for $q = 3.2$. In the first column we find values $y_n, y_{n+2}, y_{n+4}, \ldots$ converging to 2.558258; in the second column, we see $y_{n+1}, y_{n+3}, y_{n+5}, \ldots$ converging to 1.641742. Not only does the steady state s become unstable, but the y_i are attracted to a new steady *cycle of period* 2. Steady means there are two values, called *attractors*, denoted here by y_+ and y_-, such that if $y_n = y_+$, then $y_{n+1} = y_-$ and $y_{n+2} = y_+$. Clearly then

$$y_{n+j} = y_+ \quad \text{when } j \text{ is even,}$$

$$y_{n+j} = y_- \quad \text{when } j \text{ is odd.}$$

TABLE 2.1

For $q = 3.200$ we obtain a 2-cycle. The attractors are
$y_+ = 2.558258$, $y_- = 1.641742$ (to 6 places).

y_n	y_{n+1}
2.400000	1.920000
2.457600	1.824522
2.509590	1.732646
2.542405	1.671873
2.554834	1.648292
2.557668	1.642872
2.558162	1.641926
2.558242	1.641772
2.558255	1.641747
2.558257	1.641743
2.558258	1.641743
2.558258	1.641742
2.558258	1.641742
2.558258	1.641742
2.558258	1.641742
2.558258	1.641742
2.558258	1.641742
2.558258	1.641742
2.558258	1.641742

The y_i tend to this cycle in the sense that regardless of initial size successive population sizes alternately approach y_+ and y_-.

We demonstrate the existence of the steady cycle of period 2 by relating y_{n+2} to y_n directly. To do this succinctly we employ functional notation, that is we write the relation (2.28″) between y_n and y_{n+1} as

$$y_{n+1} = T(y_n),$$

where

(2.30) $$T(y) = y(q - y).$$

Then $y_{n+2} = T(T(y_n))$; so to find a y_n such that $y_{n+2} = y_n$, we have to solve the equation

(2.31) $$T(T(y)) = y.$$

A function $F(x)$ is said to have a fixed point if there is a value $x = z$ in its domain such that $F(z) = z$; this value z is called a *fixed point* of the function F. Thus a value y that satisfies (2.31) is called a *fixed point* of $T(T(y))$.

Now we sketch an algebraic solution of (2.31). Using the definition (2.30) of the function T, we write (2.31) in the form

$$T[T(y)] = T[y(q-y)] = y$$

or

$$y(q-y)[q - y(q-y)] = y$$

which is equivalent to

(2.32) $$y^4 - 2qy^3 + q(q+1)y^2 - (q^2 - 1)y = 0.$$

This is a fourth degree equation in y. But since a fixed point of $T(y)$ certainly is a fixed point of $T(T(y))$, and since $y = 0$ and $y = s = q - 1$ are known fixed points of T, we know two of the roots of (2.32). By the factor theorem we know that $y - s$ and y are factors of the fourth degree polynomial in (2.32), so we can determine the remaining quadratic factor by long division. Thus we arrive at the quadratic equation

$$y^2 - (q+1)y + q + 1 = 0,$$

whose roots

$$y_\pm = \frac{q+1}{2} \pm \sqrt{\left(\frac{q-1}{2}\right)^2 - 1}$$

are the new fixed points of $T(T(y))$. These are real and distinct precisely when $q > 3$. It is an easy exercise in algebra to verify that

$$T(y_-) = y_+, \qquad T(y_+) = y_-;$$

therefore, if $y_n = y_+$, then $y_{n+1} = y_-$, $y_{n+2} = y_+$, and so on cyclically.

What happens if y_n is close to, but not equal to y_+? Will $y_{n+2} = T[T(y)]$ be closer to y_+ than y_n was? If so, the subsequence y_n, y_{n+2}, y_{n+4} tends to y_+. Since $T(y_k) = y_{k+1}$, and since $T(y_+) = y_-$, it follows from the continuity of T that the complementary subsequence $y_{n+1}, y_{n+3}, \ldots$ tends to y_-; in other words, the cycle y_-, y_+ is stable, as Table 2.1 indicates.

We shall settle this stability question with the help of a stability criterion for fixed points. We shall need some notions of calculus.

A fixed point z of a function F (see definition on p. 53) is called *stable* if, for x close to z, $F(x)$ is closer to z than x is to z; i.e., if $|F(x) - z| < |x - z|$, or

$$\left|\frac{F(x) - z}{x - z}\right| < 1.$$

If $F(x)$ is farther from z than x is, then z is an unstable fixed point and the opposite inequality holds. Since $F(z) = z$, the quotient

$$\frac{F(x) - z}{x - z}$$

is the difference quotient

$$\frac{F(x) - F(z)}{x - z}.$$

For a differentiable function F, this quotient is close to the derivative F' of F at z when x is close to z. So, to test the stability of a fixed point z, we use this principle:

If $|F'(x)| < 1$, *then the fixed point z is stable*;
if $|F'(z)| > 1$, *then the fixed point z is unstable*.

Now reconsider the function $T(y) = y(q - y)$ defined in (2.30). Its fixed points are $y = 0$ and $y = q - 1 = s$. Let us test their stability with our newly found criterion.

GROWTH FUNCTIONS

We have $T'(y) = q - 2y$; so $T'(0) = q$ and $|T'(0)| > 1$ for all values of $q > 1$, that is, for all q relevant to our model. Therefore 0 is an unstable fixed point of $T(y)$. At the fixed point s,

$$T'(s) = q - 2s = q - 2(q - 1) = 2 - q;$$

$$|T'(s)| < 1 \quad \text{for} \quad -1 < 2 - q < 1,$$

i.e., for $1 < q < 3$. We conclude that s is a stable fixed point in the interval $1 < q < 3$. These results confirm our earlier conclusions based on the difference quotient (2.29) instead of the derivative.

Next we apply our stability criterion to the fixed points y_+ and y_- of the function $G(y) = T[T(y)]$. We compute the derivative of G by the chain rule and find $G'(y) = T'[T(y)]T'(y)$. Its values at y_+ and y_- are

$$G'(y_+) = T'[T(y_+)] \cdot T'(y_+) \quad \text{and} \quad G'(y_-) = T'[T(y_-)] \cdot T'(y_-);$$

and since $T(y_+) = y_-$ and $T(y_-) = y_+$,

$$G'(y_+) = T'(y_-)T'(y_+) = G'(y_-) = (q - 2y_+)(q - 2y_-).$$

Substituting $q + 1 \pm 2\sqrt{D}$ for $2y_\pm$, where $D = \tfrac{1}{4}(q-1)^2 - 1$, see p. 54, we find

$$G'(y_\pm) = (-1 - 2\sqrt{D})(-1 + 2\sqrt{D}) = 1 - 4D = 5 - (q-1)^2.$$

Note that

$$|G'(y_\pm)| < 1 \quad \text{for} \quad -1 < (q-1)^2 - 5 < 1.$$

These inequalities for q are equivalent to $2 < q - 1 < \sqrt{6}$, and to

$$3 < q < 1 + \sqrt{6} \approx 3.44949.$$

For each q in this interval, there is a stable cycle of period 2.

Higher values of q rarely occur in practical applications of Verhulst's law, but they do figure importantly in other applications[†] of equation (2.28). We therefore describe, without proofs, this behavior of $\{y_n\}$.

When q exceeds $1 + \sqrt{6}$, the behavior of the sequence $\{y_n\}$ becomes still more complicated. Table 2.2 was generated for $q = 3.52 > 1 + \sqrt{6}$

[†]E.g. "Universal Behavior in Nonlinear Systems" by Mitchell J. Feigenbaum, *Los Alamos Science*, Summer 1980.

TABLE 2.2

For $q = 3.520$ we obtain a 4-cycle. The attractors are
3.095793, 1.313257, 2.898021, 1.802509 (to 6 places).

y_n	y_{n+1}	y_{n+2}	y_{n+3}
3.040000	1.459200	3.007119	1.542293
3.050204	1.432974	2.990654	1.583090
3.066303	1.391172	2.961566	1.653839
3.086330	1.338449	2.919895	1.752243
3.097540	1.308587	2.893826	1.812038
3.094892	1.315663	2.900165	1.797624
3.096184	1.312211	2.897085	1.804637
3.095608	1.313753	2.898463	1.801502
3.095878	1.313031	2.897819	1.802969
3.095754	1.313362	2.898115	1.802295
3.095811	1.313209	2.897978	1.802607
3.095785	1.313279	2.898041	1.802463
3.095797	1.313247	2.898012	1.802530
3.095791	1.313262	2.898025	1.802499
3.095794	1.313255	2.898019	1.802513
3.095793	1.313258	2.898022	1.802507
3.095793	1.313257	2.898020	1.802510
3.095793	1.313257	2.898021	1.802508
3.095793	1.313257	2.898021	1.802509
3.095793	1.313257	2.898021	1.802509

and shows a new cycle of period 4. Each row lists consecutive values of y_i, so that each column lists every fourth iterate: y_n, y_{n+4}, y_{n+8}, ... Each of these subsequences converges to one of the four listed "attractors". Indeed for $1 + \sqrt{6} < q < q_4$ (where q_4 is the upper end point of the parameter interval that produces 4-cycles), a cycle of period 4 takes over the role of attracting the y_n, while the previous cycle of period 2 loses its stability.

For $q_4 < q < q_5$, a stable cycle of period 8 appears, and so on. For $q_k < q < q_{k+1}$, a cycle of period 2^{k-1} appears. Each new cycle has twice the period of the previous and is stable; the y_n get closer and closer to it. In each interval $q_k < q < q_{k+1}$, only the cycle with largest period, 2^{k-1}, is stable. How do these critical parameter values q_k behave?

The values q_k where new cycles emerge satisfy

$$q_1 < q_2 < \cdots < q^*, \qquad \lim_{k \to \infty} q_k = q^* < 4.$$

For $q^* < q < 4$, the behavior of the sequence of population sizes is so complicated that it is called *chaotic*.

GROWTH FUNCTIONS

Is Verhulst's formula (2.27) reliable? Around about 1850 he made a careful population study of several European countries and of the United States. He used his law to predict their populations as far ahead as a century. Some of his predictions are famous, and justly so. For example, he calculated that France would reach a maximum population of 40 million in 1921; the event proved him correct. Despite the Civil War (1861–65) his prediction for the U.S. population in 1940 was off by less than a million. But ironically, his law applied to his own country, Belgium, did not work. Belgium's population curve for the century is given by the solid graph in Figure 2.10.

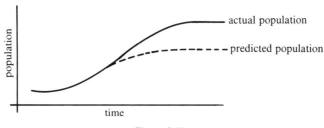

Figure 2.10

How did Verhulst's prediction go wrong? Perhaps because Belgium switched from agriculture to industry and colonized the Congo. This distinct sociological change *permanently altered* the growth and competition coefficients. His application of his law continued to describe the growth of Belgium as agricultural when it was in fact industrial. Observe that the Belgian population curve is a combination of the parts of two S-curves, the earlier with agricultural, and the later with industrial, conditions obtaining. How then, it may well be asked, was his prediction for the United States successful despite the Civil War? Perhaps because Verhulst could not know that the Civil War was going to break out a decade or so after he made his population analysis. Even though the U.S. was also shifting from an agricultural to an industrial society, he could not take the changed values of the growth and competition coefficients due to the war into account. The point is that these changes of coefficients, unlike those due to a switch from agriculture to industry and colonization in Belgium, were merely *temporary*: soon after the Civil War his coefficients were again accurately descriptive. With people killed in the war his 1865 population estimate was too high, but both his estimate and the actual 1865 population have growths asymptotic to $(q - 1)/r$. In the long run his prediction would have been correct; the run to 1940 was long enough for it to be correct within one million. Verhulst could not have foreseen the impact of science and technology (health care, birth control, the green revolution) on his coefficients.

Although quite elementary, the investigation of Verhulst's law (2.27) was rather lengthy; one therefore often prefers an approximate approach via differential equations which we outline here for readers who have had a first course in calculus.

A differential equation involves the derivative of a function; so instead of dealing with a function $y(n) = y_n$ defined only on the integers, we need a differentiable function $y(t)$ of time defined for all real numbers t. If we take our equation (2.28) and subtract y_n from both sides, we can write it in the form of a difference equation

(2.28d) $$y_{n+1} - y_n = (q-1)y_n - y_n^2,$$

or, equivalently, as

(2.28d′) $$\frac{y(n+1) - y(n)}{n+1-n} = (q-1)y_n - y_n^2.$$

Now suppose there is a differentiable function which, for integer inputs $t = n$, has values $y(n) = y_n$; if the difference quotient on the left of (2.28d′) is well approximated by the derivative dy/dt at t, we might hope that the solution of the differential equation

(2.28d″) $$\frac{dy}{dt} = (q-1)y - y^2$$

at $t = n$ approximates y_n.

It is easy to solve this equation for any given initial value $y(0)$; we leave it to readers who have had a first course in calculus to verify that the function

$$y(t) = \frac{q-1}{Ae^{-(q-1)t} + 1},$$

where A is a constant related to the initial value $y(0)$ by

$$A = \frac{q-1}{y(0)} - 1,$$

solves (2.28d″) for all t. For *any* $q > 1$, this function tends to $(q-1)$ as $t \to \infty$. Note that this corresponds to the steady state $s = q - 1$ obtained in the discrete approach. But there, as soon as q exceeds 2, this state is unstable and the population size vacillates; whereas the solution of the differential equation tends to the steady state $q - 1$, no matter how large the parameter q is. Thus, for $q > 2$, the solutions of the differential equation do *not* approximate those of the difference equation, and our precise method confirms a phenomenon that has been observed but cannot be deduced by the idealization via calculus.

2.5 Cusanus' Recursive Formula for π

When, as in the last section, a member x_n of a sequence is defined in terms of earlier members of the sequence, it is said to be defined *recursively*. This terminology emphasizes the fact that the sequence refers back to itself; it is, so to speak, a snake biting its own tail.

We now consider one of the most elegant recursive formulae in mathematics, namely that given by Cusanus (1401–64) in about 1450. Even though it was the first to facilitate a more accurate calculation of π than that given by Archimedes' approximations, it is not widely known. More than five hundred years old, it is perhaps too modern for the "modernists". With this formula we have a hint that there was, contrary to popular historical misconception, tremendous intellectual activity before the Sixteenth Century. Despite what the history books fail to say, without Cusanus and his ilk, Galileo and Newton could not have inherited the groundwork they did in fact inherit.

Cusanus' calculation of π. It really is obvious that if a regular polygon of perimeter p is circumscribed by a circle of radius R, then the more sides the polygon has, the closer the approximation $p/2R$ to π. Surely thousands of persons before and after Archimedes must have thought of this, yet how many have found a method of effectively exploiting it to calculate π? Archimedes considered an unending sequence of regular polygons, each polygon with more sides than its predecessor, each circumscribed by the same circle; Cusanus considered an unending sequence of regular polygons, each polygon with more sides than its predecessor, but all of the same perimeter and therefore circumscribed by different circles. Whereas Archimedes found the limit of p with constant R, Cusanus found the limit of R with constant p. Both methods are elegant. Those who find Cusanus' elegance exciting may enjoy studying Archimedes' derivation on their own.

How, specifically, did Cusanus exploit his idea? He did so in the following way. From a given circle C_1 of radius r_1 circumscribing a regular polygon of m sides and perimeter k, another circle C_2 of radius r_2 circumscribing a regular polygon of $2m$ sides, but with the same perimeter, is constructed. By repetition of the procedure n times there results a sequence of circles $C_1, C_2, C_3, \ldots, C_{n+1}$, of radii $r_1, r_2, r_3, \ldots, r_{n+1}$, circumscribing regular polygons *with constant perimeter k*, of m, $2m$, $2^2 m, \ldots, 2^n m$ sides, respectively. It is intuitively clear that

$$\pi = \frac{k}{2R}, \qquad \text{where} \quad R = \lim_{n \to \infty} r_{n+1}.$$

(Since we are now considering a sequence of polygons of constant perimeter, we use the letter k in preference to p.) The real problem, of course, is to determine r_{n+1}. The way in which C_2 is constructed from C_1 determines the relation between r_2 and r_1. But C_3 is constructed from C_2 as C_2 from C_1, so that r_3 has the same relation to r_2 as r_2 to r_1, and for similar reasons r_4 has the same relation to r_3 as r_3 has to r_2. Thus r_4 can be determined in terms of r_3, while r_3 can be determined in terms of r_2, and r_2 in terms of r_1, so that finally r_4 can be determined in terms of r_1. More

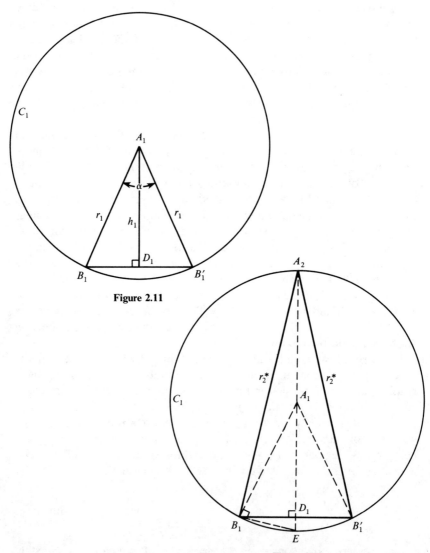

Figure 2.11

Figure 2.12

GROWTH FUNCTIONS

generally, r_{n+1} is determined in terms of r_n, which in turn is determined in terms of r_{n-1}, etc., so that finally r_{n+1} is determined in terms of r_1. The formula is recursive.

Now for the details. What, specifically, is the relation between r_2 and r_1? Figure 2.11 illustrates the essentials of what we are given: the m-sided circumscribed polygon of perimeter k being regular, it is sufficient to consider just one of its sides. We make the construction illustrated by Figure 2.12.

Since, as Euclid tells us, angle subtended at circumference is one-half angle subtended at center,

$$\angle B_1 A_2 B_1' = \tfrac{1}{2} \angle B_1 A_1 B_1'.$$

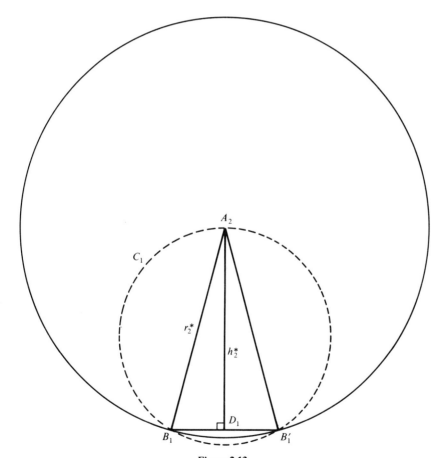

Figure 2.13

Consequently $2m$ such triangles as $B_1A_2B_1'$ fit together to form a regular polygon with perimeter $2m\overline{B_1B_1'}$, which is circumscribable by a circle C* with center A_2 and (say) radius r_2^*. Figure 2.13 illustrates the essentials. Retaining A_2 as center we now shrink Figure 2.13 to half size. We then have a circle C_2 circumscribing a regular polygon with perimeter k, but with $2m$ sides. Compare Figure 2.14 with Figure 2.11.

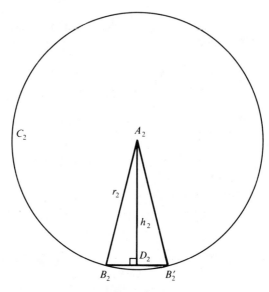

Figure 2.14

The problem is to find r_2 in terms of r_1. To do this we first consider the geometry of Figure 2.12. $\angle A_2B_1E$ is a right angle, since it is inscribed in a semicircle. Thus triangles A_2B_1E, $A_2B_1D_1$ are both right triangles (the latter is right angled at D_1) and additionally have a common angle, $\angle B_1A_2E$. Therefore these triangles are similar, and consequently their corresponding sides are proportional. Considering the sides opposite the equal angles $\angle A_2EB_1, \angle A_2B_1D_1$ and the sides opposite the right angles $\angle A_2B_1E, \angle A_2D_1B_1$, we therefore have

$$\frac{A_2B_1}{A_2D_1} = \frac{A_2E}{A_2B_1},$$

i.e.,

$$\frac{r_2^*}{h_2^*} = \frac{2r_1}{r_2^*}, \qquad \text{where } h_2^* = A_2D_1.$$

GROWTH FUNCTIONS

Thus

(2.33) $$\left(r_2^*\right)^2 = 2r_1 \cdot h_2^*.$$

But h_2^* is an uninvited bedfellow and is speedily to be replaced. We have

$$h_2^* = A_2 D_1 = A_2 A_1 + A_1 D_1,$$

i.e.

(2.34) $$h_2^* = r_1 + h_1.$$

We have related the measurements of Figure 2.11 to those of Figure 2.13; we wish to relate them to those of Figure 2.14. But Figure 2.14 was obtained from Figure 2.13 by reducing everything to half size, so that

$$r_2 = \tfrac{1}{2} r_2^*, \qquad h_2 = \tfrac{1}{2} h_2^*.$$

Hence from (2.34) we have

(2.35) $$h_2 = \frac{r_1 + h_1}{2},$$

and from (2.33)

$$r_2^2 = \tfrac{1}{4}\left(r_2^*\right)^2 = \tfrac{1}{4} 2 r_1 2 h_2,$$

so that

(2.36) $$r_2 = \sqrt{r_1 h_2}.$$

This derivation discloses our motive for using a star notation: to emphasize the transitory roles of r_2^* and h_2^*.

Equations (2.35) and (2.36) give r_2 in terms of r_1 (and h_1). The intrusion of the h's is an incidental complexity that must not be permitted to obscure the leading idea; in specifying the relation between r_2 and r_1 (and h_1) we have reached the heart of the matter. In repeating our procedure to obtain C_3 from C_2 as C_2 was obtained from C_1, r_3 will have the same relation to r_2 as r_2 has to r_1, and in obtaining C_4 from C_3, r_4 will have the same relation to r_3 as r_3 has to r_2 and r_2 has to r_1. Consequently, for C_{n+1} we have

(2.37) $$h_{n+1} = \frac{r_n + h_n}{2}.$$

and

$$r_{n+1} = \sqrt{r_n h_{n+1}}. \tag{2.38}$$

Let us recapitulate. To avoid the verbosity of saying that h_n is the altitude of any triangle whose vertex is the center of C_n and whose base is one of the sides of the regular polygon of $2^{n-1}m$ sides circumscribed by C_n, let us refer to h_n as the apothem of C_n. Accordingly, if C_1 is a circle of radius r_1 and apothem h_1 circumscribing a regular polygon of m sides and perimeter k, then by repeating n times the process considered above we form a sequence of circles $C_1, C_2, C_3, \ldots, C_{n+1}$ having radii $r_1, r_2, r_3, \ldots, r_{n+1}$ (and apothems $h_1, h_2, h_3, \ldots, h_{n+1}$), circumscribing regular polygons of $m, 2m, 2^2m, \ldots, 2^n m$ sides, respectively, where h_{n+1}, r_{n+1} satisfy (2.37), (2.38) for $n = 0, 1, 2, \ldots$.

To calculate π, i.e., $k/2R$, it merely remains to determine R, where $R = \lim_{n \to \infty} r_{n+1}$. It is convenient to take C_1 as circumscribing a regular hexagon, i.e., to take $m = 6$, and to take $r_1 = 1$. See Fig. 2.15. Here h_1 is evidently the altitude of an equilateral triangle of unit side. By simple calculation we find $h_1 = \sqrt{3}/2$. The reader is now in a position to calculate a sequence of successively better approximations to R, and hence, to π.

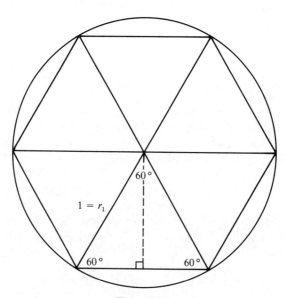

Figure 2.15

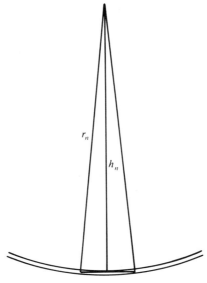

Figure 2.16

This raises the question of the accuracy of approximations. It is obvious that, as n increases, the angle at the vertex of each triangle constituting the regular polygon circumscribed by C_n approaches zero, and therefore h_n/r_n approaches 1. Thus

$$\lim_{n \to \infty} r_n = \lim_{n \to \infty} h_n,$$

i.e., r_n and h_n both converge to R. And since the hypotenuse is the greatest side of a right triangle,

$$r_n > h_n;$$

see Figure 2.16. Consider the polygon; h_n is the radius of its inscribed circle and r_n is the radius of its circumscribed circle; in the limit these circles coincide. Thus it should be clear that r_n decreases and h_n increases as n increases, and that

$$r_n > R > h_n.$$

The astonishing thing is that we are able to anticipate that, for example, for the hexagon, with $r_1 = 1$, and consequently with perimeter $k = 6$, the repeated applications of (2.37) and (2.38) will converge to $6/2\pi$, i.e, to $3/\pi$.

We may add that neither Cusanus nor Descartes (1596–1650) (who made extensive use of Cusanus' formulae) worried overmuch about convergence: they were confident of their intuition.

With no more than an elementary knowledge of inequalities we can prove convergence. The crux of the matter is that the difference between r_n and h_n gets smaller and smaller. But having the abhorrence for square roots that Pythagoras had for eating beans, we prefer to consider the difference of r_n^2 and h_n^2. By (2.37), (2.38),

$$r_{n+1}^2 - h_{n+1}^2 = r_n h_{n+1} - \left(\frac{r_n + h_n}{2}\right)^2;$$

by (2.37) the product of r_n and h_{n+1} is

$$r_n(h_{n+1}) = r_n\left(\frac{r_n + h_n}{2}\right),$$

so

$$\begin{aligned} r_{n+1}^2 - h_{n+1}^2 &= r_n\left(\frac{r_n + h_n}{2}\right) - \left(\frac{r_n + h_n}{2}\right)^2 \\ &= \left(\frac{r_n + h_n}{2}\right)\left(r_n - \frac{r_n + h_n}{2}\right) \\ &= \left(\frac{r_n + h_n}{2}\right)\left(\frac{r_n - h_n}{2}\right) \\ &= \tfrac{1}{4}\left(r_n^2 - h_n^2\right). \end{aligned}$$

Thus

$$r_3^2 - h_3^2 = \tfrac{1}{4}(r_2^2 - h_2^2), \quad \text{and} \quad r_2^2 - h_2^2 = \tfrac{1}{4}(r_1^2 - h_1^2),$$

so that

$$r_3^2 - h_3^2 = \frac{1}{4^2}(r_1^2 - h_1^2).$$

Proceeding in this way, after n steps, we have

(2.39) $$r_{n+1}^2 - h_{n+1}^2 = \frac{1}{4^n}(r_1^2 - h_1^2).$$

GROWTH FUNCTIONS

But the hypotenuse r_1 of the triangle is greater than its leg h_1 (see Figure 2.11), so that $r_1^2 - h_1^2$ is positive and hence

(2.40) $$r_{n+1} > h_{n+1}.$$

Thus $r_n - h_n > 0$ for $n = 1, 2, \ldots$. By (2.37),

$$h_{n+1} - h_n = \tfrac{1}{2}(r_n - h_n),$$

and therefore

(2.41) $$h_{n+1} > h_n.$$

Squaring (2.38) and dividing by $r_n r_{n+1}$, we have

$$\frac{r_{n+1}}{r_n} = \frac{h_{n+1}}{r_{n+1}}$$

which, by (2.40), is less than 1 so that

$$\frac{r_{n+1}}{r_n} < 1,$$

and therefore

(2.42) $$r_{n+1} < r_n.$$

Algebra together with obvious geometry confirms intuition. Inequality (2.41) shows that successive values of h_n increase (thus if $\lim_{n \to \infty} h_n = R$, then $h_n < R$), while (2.42) shows that successive values of r_n decrease (so that if $\lim r_n = R$, then $R < r_n$), and (2.39) shows that if either r_n or h_n converges then both converge to the same limit since their difference converges to zero. Together these results imply that there is an R such that for all n, $h_n < R < r_n$, and that

$$\lim_{n \to \infty} h_n = R = \lim_{n \to \infty} r_n.$$

Let's be specific. Taking the hexagon as our initial polygon, with $r_1 = 1$ and (consequently) $h_1 = \sqrt{3}/2$, (2.39) yields

$$r_{n+1}^2 - h_{n+1}^2 = \frac{1}{4^n}\left(1^2 - \frac{3}{4}\right) = \frac{1}{4^{n+1}}$$

so that

(2.43) $$r_{n+1} - h_{n+1} = \frac{1}{4^{n+1}} \cdot \frac{1}{r_{n+1} + h_{n+1}}.$$

But (2.41) implies that $h_{n+1} > h_1 = \sqrt{3}/2$, so by (2.40), we have $r_{n+1} > h_{n+1} > \sqrt{3}/2$; hence

$$r_{n+1} + h_{n+1} > \frac{\sqrt{3}}{2} + \frac{\sqrt{3}}{2}.$$

Its reciprocal therefore satisfies

$$\frac{1}{r_{n+1} + h_{n+1}} < \frac{1}{\sqrt{3}} < 1,$$

and equation (2.43) yields the inequality

$$r_{n+1} - h_{n+1} < \frac{1}{4^{n+1}} \cdot \frac{1}{\sqrt{3}} < \frac{1}{4^{n+1}}.$$

This says that the difference between the radii of the circumscribed and inscribed circles of the polygon obtained from the initial hexagon after n steps is $< 1/4^{n+1}$; convergence is rapid.

Since $2\pi R = k$, and in this example the perimeter k of the hexagon is 6, we have

$$R = \frac{6}{2\pi} = \frac{3}{\pi}$$

(as noted earlier), so that the successive values of h_n increase to $3/\pi$ while the successive values of r_n decrease to it. That is,

$$h_n < \frac{3}{\pi} < r_n,$$

whence

$$\frac{1}{r_n} < \frac{\pi}{3} < \frac{1}{h_n} \quad \text{and} \quad \frac{3}{r_n} < \pi < \frac{3}{h_n}.$$

Even the case $n = 1$ is interesting:

$$\frac{3}{1} < \frac{3}{\sqrt{3}/2} = 2\sqrt{3}, \quad \text{i.e., } 3 < \pi < 2\sqrt{3} \approx 3.4.$$

Surely the reader will want to work out $n = 2, 3, 4$ (and maybe others) for himself.

Finally, with the suggestion that the reader take a second look at Figure 2.16 and the reminder that

$$\lim_{\theta \to 0} \frac{\sin \theta}{\theta} = 1,$$

he is urged to prove that, in the general case, the common limit R of r_n and h_n is given by

$$R = \frac{\sqrt{r_1^2 - h_1^2}}{\arccos(h_1/r_1)}.$$

2.6 Arithmetic and Geometric Means

M_A, the arithmetic mean of n non-negative numbers $a_1, a_2, a_3, \ldots, a_n$, is defined by

$$M_A = \frac{a_1 + a_2 + a_3 + \cdots + a_n}{n}.$$

M_G, the geometric mean of these quantities, is defined by

$$M_G = \sqrt[n]{a_1 \cdot a_2 \cdot a_3 \cdots a_n}.$$

Thus, for example, (2.37) of the last section states that h_{n+1} is the arithmetic mean of r_n and h_n, while (2.38) states that r_{n+1} is the geometric mean of r_n and h_{n+1}.

If the quantities $a_1, a_2, a_3, \ldots,$ are all equal, then

$$M_A = \frac{na_1}{n} = a_1 \quad \text{and} \quad M_G = \sqrt[n]{a_1^n} = a_1,$$

so that $M_A = M_G$. If the quantities are not all equal, then $M_A > M_G$. This is very easily proved in the simple case $n = 2$. For the two quantities a, b we have

$$M_A = \frac{a+b}{2}, \qquad M_G = \sqrt{ab}.$$

Therefore

$$M_A - M_G = \tfrac{1}{2}(a + b - 2\sqrt{ab}) = \tfrac{1}{2}\left[(\sqrt{a})^2 - 2\sqrt{a}\sqrt{b} + (\sqrt{b})^2\right]$$
$$= \tfrac{1}{2}(\sqrt{a} - \sqrt{b})^2;$$

but $\tfrac{1}{2}(\sqrt{a} - \sqrt{b})^2 > 0$ unless $a = b$. This proves the proposition for $n = 2$. The inequality

(2.44) $$M_A \geq M_G,$$

for n non-negative numbers $a_1, a_2, \ldots, a_n$ is of great importance in many fields of pure and applied mathematics. The great French mathematician Augustin Louis Cauchy (1789–1857) gave such an ingenious proof for it that we cannot resist the temptation to include it here. It is based on a curious method of induction. We have just established it for two numbers, say a_1 and a_2. Consider now four numbers a_1, a_2, a_3 and a_4. Let

$$s_1 = \frac{a_1 + a_2}{2}, \qquad s_2 = \frac{a_3 + a_4}{2}.$$

Then by our result for two numbers,

$$s_1 \geq \sqrt{a_1 a_2}, \qquad s_2 \geq \sqrt{a_3 a_4}.$$

Now we use the inequality involving two numbers again and find

$$\frac{s_1 + s_2}{2} \geq \sqrt{s_1 s_2},$$

or in terms of the a's

$$\frac{1}{2}\left[\frac{a_1 + a_2}{2} + \frac{a_3 + a_4}{2}\right] \geq \sqrt{\frac{a_1 + a_2}{2} \frac{a_3 + a_4}{2}}$$
$$\geq \sqrt{\sqrt{a_1 a_2} \sqrt{a_3 a_4}}.$$

But this just means that

$$M_A = \frac{a_1 + a_2 + a_3 + a_4}{4} \geq \sqrt[4]{a_1 a_2 a_3 a_4} = M_G$$

and proves the inequality for $n = 4$. We repeat the argument for $n = 8$.

Indeed, let

$$t_1 = \frac{a_1 + a_2 + a_3 + a_4}{4}, \quad t_2 = \frac{a_5 + a_6 + a_7 + a_8}{4}.$$

Now we know by our preceding result that

$$t_1 \geq \sqrt[4]{a_1 a_2 a_3 a_4} \quad \text{and} \quad t_2 \geq \sqrt[4]{a_5 a_6 a_7 a_8}.$$

Hence by the inequality for $n = 2$ we obtain

$$M_A = \tfrac{1}{8}(a_1 + a_2 + a_3 + a_4 + a_5 + a_6 + a_7 + a_8)$$

$$= \tfrac{1}{2}(t_1 + t_2) \geq \sqrt{t_1 t_2} = \sqrt[8]{a_1 a_2 a_3 a_4 a_5 a_6 a_7 a_8} = M_G.$$

We can obviously continue the game indefinitely and find

$$M_A \geq M_G$$

if we deal with $n = 2^l$ positive numbers a_k.

The last step consists of getting rid of the restriction that n is a power of 2. We take any n and find the integer l such that

$$2^{l-1} < n \leq 2^l.$$

We complete the set of n numbers $a_1, \ldots, a_n$ by $2^l - n$ constants a_0. We now have 2^l numbers (of which the last $2^l - n$ all have the value a_0) and apply the inequality for 2^l numbers,

$$\frac{1}{2^l}[a_1 + a_2 + \cdots + a_n + (2^l - n)a_0] \geq \left[a_1 a_2 \cdots a_n a_0^{2^l - n}\right]^{1/2^l}.$$

Since $a_1 + a_2 + \cdots + a_n = nM_A$, and $a_1 a_2 \cdots a_n = M_G^n$, this leads to

$$\frac{1}{2^l}[nM_A + (2^l - n)a_0] \geq M_G^{n/2^l} a_0^{1 - (n/2^l)}.$$

Setting $a_0 = M_A$ in this inequality, we obtain

$$M_A \geq M_G^{n/2^l} M_A^{1 - (n/2^l)}, \quad M_A^{n/2^l} \geq M_G^{n/2^l}$$

and thus finally the desired result

$$M_A \geq M_G$$

for any n non-negative numbers.

What are the uses of these means? We saw already in Section 2.2 that the arithmetic mean is used to obtain the best possible information from n repeated measurements of a quantity x. This arose in connection with Gauss' principle of least squares. Less well known is his application of the geometric mean which follows.

How is weight W to be accurately determined by using badly made scales? How, for example, with scales of which one arm is longer than the other? We suppose that W, when placed in the left and right pans, counterbalances weights W_1, W_2, respectively. What is the actual weight of W? Study Figures 2.17 and 2.18. For equilibrium in Figure 2.17, we require

(2.45) $$l_2 W = l_1 W_1,$$

and for equilibrium in Figure 2.18, we require

(2.46) $$l_1 W = l_2 W_2.$$

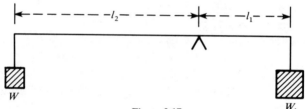

Figure 2.17

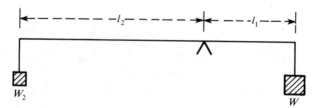

Figure 2.18

Now we come to Gauss' important observation. The product of these two equations yields

$$l_1 l_2 W^2 = l_1 l_2 W_1 W_2,$$

so that

$$W = \sqrt{W_1 W_2}.$$

GROWTH FUNCTIONS

Thus W turns out to be the geometric mean of W_1 and W_2 and is independent of the lengths of the arms. Use of the geometric mean rectifies this imprecision of the scales.

In Section 2.1 we used computations and our financial intuition to conclude that the sequence of numbers $A_n = (1 + 1/n)^n$ increases as n increases, yet remains bounded. We promised to lead the readers through exercises whose results constitute a mathematical proof of relation (2.12) on p. 32. We postponed the matter until now because we make heavy use of the arithmetic-geometric mean inequality $M_A \geq M_G$ established in this section for the means of arbitrarily many positive numbers.

In Exercise 2.3 we ask the readers to use this inequality in order to show that

$$\left(1 + \frac{1}{n}\right)^n < \left(1 + \frac{1}{n+1}\right)^{n+1}, \quad \text{i.e.} \quad A_n < A_{n+1}$$

for $n = 1, 2, \ldots$.

After they have established the increasing nature of the sequence A_n, we ask them to examine another sequence defined as

$$B_n = \left(1 + \frac{1}{n}\right)^{n+1}$$

to show that $A_n < B_n$ for $n = 1, 2, \ldots$ (in Exercise 2.4) and to apply the inequality $M_A \geq M_G$ again to demonstrate that the B_n decrease as n increases:

$$\left(1 + \frac{1}{n}\right)^{n+1} > \left(1 + \frac{1}{n+1}\right)^{n+2}, \quad \text{i.e.} \quad B_n > B_{n+1}$$

for $n = 1, 2, \ldots$.

(see Exercise 2.5).

At this point we shall be able to conclude that the A_n *increase* to a limiting value, and the B_n *decrease* to a limiting value; the point of Exercises 2.6 and 2.7 is to show that A_n and B_n tend to the same limit, called e.

The proof of $\lim_{n \to \infty} A_n = e$ outlined here is a powerful application of the inequality $M_A \geq M_G$ and uses nothing more sophisticated than algebra and properties of real numbers.

Exercise 2.3. Prove that $A_n < A_{n+1}$, i.e. that

$$\left(1 + \frac{1}{n}\right)^n < \left(1 + \frac{1}{n+1}\right)^{n+1} \quad \text{for } n = 1, 2, \ldots.$$

[Hint: Compute the arithmetic and geometric means M_A and M_G of the $n + 1$ positive numbers

$$a_1 = a_2 = \cdots = a_n = 1 + \frac{1}{n}, \quad a_{n+1} = 1,$$

and use $M_A > M_G$.]

Exercise 2.4. Show that $A_n < B_n$, i.e. that

$$\left(1 + \frac{1}{n}\right)^n < \left(1 + \frac{1}{n}\right)^{n+1}, \quad \text{for } n = 1, 2, \ldots.$$

Exercise 2.5. Prove that $B_n > B_{n+1}$, i.e. that

$$\left(1 + \frac{1}{n}\right)^{n+1} > \left(1 + \frac{1}{n+1}\right)^{n+2} \quad \text{for } n = 1, 2, \ldots.$$

[Hint: Prove the equivalent inequality $1/B_n < 1/B_{n+1}$ by applying $M_G < M_A$ to the means of the $n + 2$ positive numbers

$$b_1 = b_2 = \cdots = b_{n+1} = \frac{n}{n+1}, \quad b_{n+2} = 1].$$

Exercise 2.6. Show for $n = 1, 2, \ldots$ that $B_n = A_n(1 + 1/n)$, so that $B_n - A_n = A_n/n$; use the results of Exercises 2.3 and 2.4 to show that $A_n < B_n < B_1 = 4$, and conclude that

$$0 < B_n - A_n < \frac{4}{n} \to 0 \text{ as } n \to \infty.$$

Exercise 2.7. Picture the intervals $I_n = [A_n, B_n]$ (with left endpoints A_n, right endpoints B_n) on the number line. Combine results of previous exercises to show that
(a) I_{n+1} is contained in I_n;
(b) $\lim_{n \to \infty}$ (length of I_n) = 0.
These "nested intervals" $I_1, I_2, \ldots$ have exactly one point in common. This point is e.

CHAPTER THREE

The Role of Mathematics in Optics

To illustrate the part played by mathematics in the construction of scientific theories, we consider the development of optics.

3.1 Euclid's Optics

We begin with Euclid (c. 300 BC). Not unnaturally for a geometer, he wished, as doubtless had many geometers before him, to apply geometry to optics. Unlike the others he was successful. Conceiving light as propagated in straight lines enabled him to apply geometry to optics. On second thought this statement cannot stand. Until Euclid had applied geometry to optics there was, to use the Irish idiom, no such subject as optics. Nowadays, when diagrams are used as an ingredient of educated common sense, of course it is obvious that light is propagated in straight lines. If light rays could not be represented by lines, optical phenomena could not be illustrated by diagrams. We, with the arrogance of hindsight, cannot begin to understand Euclid's foresight in making his basic assertion that light is rectilinearly propagated. When the needle in the haystack has been pointed out to us, we are prone to suppose that finding it was no problem at all.

Physical objects that more or less crudely approximate straight lines readily come to mind, for example, a taut wire. But surely a shaft of sunlight piercing the shutters of a darkened room is singularly apt. Isn't this the perfect example? Euclid must have been well pleased with his observation. Yet note that his basic assertion embraces metaphysical speculation as well as physical observation. We see only the shafts of light at which we look; we do not see the shafts with which we look. We cannot observe the rays with which we observe, yet Euclid claims all rays to be propagated in straight lines. Such metaphysical assumptions regarding unobservables are acceptable in so far as they facilitate understanding of observables. Is his postulate obvious? Your answer depends upon how much or how little you think about it.

Given that rays of light are straight lines, how, Euclid asked, is the direction of a ray striking the surface of a plane mirror related to that of the reflected ray? See Figure 3.1. This figure reduces optics to geometry. The lines l_1, l_2, n, represent the incident ray, the reflected ray, and the normal to the surface at the point of incidence, respectively. The angle α between incident ray and normal is termed the *angle of incidence*, while the angle β between reflected ray and normal is termed the *angle of reflection*. What is the relation between β and α?

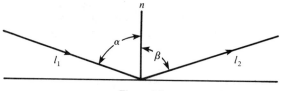

Figure 3.1

Euclid found by experiment that l_2 lies in the plane determined by l_1 and n. Thus l_1, n, and l_2 in Figure 3.1 may be considered to lie in the plane of the paper. To determine l_2 uniquely, it remains to specify β. As the result of many experiments Euclid found that $\beta = \alpha$, i.e., that angle of reflection is equal to angle of incidence. This is the famous law of reflection as formulated by him in his *Optics*.

Although this law was based on a large number of experiments we must remember that Greek technology was rudimentary, their measuring instruments imprecise, and their plane mirrors imperfect. What assurance had Euclid that β was precisely equal to α? He had the comforting security of experiment backed by belief. He held a possibly only half-articulate, but certainly deep-seated, belief about the nature of things; that Nature is not fortuitous, that her laws have simplicity and elegance. With the courage of conviction he asserted his law to hold exactly for perfectly plane mirrors. But many of Euclid's contemporaries, even if equally courageous, had grave doubts whether his law is right. Some with different metaphysics doubted if there could be laws of nature at all.

3.2 Heron: The Shortest Path Principle

To add grounds for belief we introduce Heron of Alexandria, who lived several centuries after Euclid, perhaps about 100 A.D. (His birth and death dates are uncertain.) A man who played a far greater role in

MATHEMATICS IN OPTICS

the development of science than that usually ascribed to him in the general run of textbooks, he built the first automaton, made the first attempt at building a steam engine, developed trigonometry and applied it extensively. A man with both feet on the ground, he was forever stressing the possibilities of applying mathematics.

Heron gave a proof of Euclid's law of reflection. His proof consists of showing that *both* of Euclid's laws:

E_1 Light is propagated rectilinearly,

E_2 Angle of Reflection = Angle of Incidence

are consequences of the principle proposed by Heron himself, that

H   Light takes the shortest path possible.

Here we have what is probably the first example of the unifying trend so characteristic of science. Surely either of E_1, E_2 could be true without the other. Is it not perfectly reasonable to conceive of light being propagated in straight lines without $\beta = \alpha$? But H could not be true without both E_1 and E_2 being true. Moreover, the complete formulation of E_2 is complicated, while H, like E_1, is simple. Is it not easier to believe one statement of a certain kind than twenty or two of the same or a more complicated kind? It is in this sense that Heron "proved" Euclid's law of reflection.

The proof that E_1 follows from H is obvious. Since the shortest distance between any two points A and B (in free space) is the straight line AB that joins them, light, in moving from A to B by the shortest path possible, is necessarily propagated rectilinearly. Figure 3.2 is self-explanatory.

Figure 3.2

The proof that E_2 follows from H is not obvious. We suppose light to travel from A via some point P in the mirror surface to B. PN is the normal to the mirror surface at P. See Figure 3.3. If the light did not become incident to the mirror surface, then the light could not be reflected from it. Here, in asserting that a ray takes the shortest path possible from A to B, we cannot mean the shortest of all possible paths (the straight line AB); we must mean the shortest possible path via the *surface of the mirror*. Thus to prove that E_2 is a consequence of H is to prove that, if *APB* is the shortest path possible (via the mirror surface), then the angles made by the *straight* lines *AP*, *PB* with *PN* are equal.

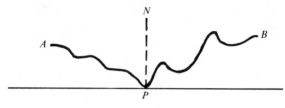

Figure 3.3

First we show that the lines *AP*, *PB* cannot be wiggly. The path from *A* to *B* via *P* will have minimum length when *AP* and *PB* both have minimum lengths, for if both were not minimal, their sum could be decreased. But the minimal path between any two points is the straight line joining them, so that the path from *A* to *B* via *P* can be minimal only if both *AP* and *PB* are straight line segments. Accordingly, we exclude wiggly paths from further consideration.

This leads us to the crux of the proof. What is the position of *P* such that the sum of the straight line distances *AP*, *PB* is a minimum? At this stage we avail ourselves of Heron's ingenuity by introducing an auxiliary point *B'*, the mirror image of *B*. That is to say, *B'* is the point on the normal from *B* to the mirror as far below the surface as *B* is above it. See Figure 3.4.

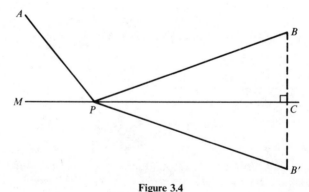

Figure 3.4

Since *MC* is perpendicular to *BB'* and *C* is the midpoint of *BB'*, *MC* is the perpendicular bisector of *BB'*; i.e. *MC* is the locus of points equidistant from B and B'. Hence for any point P on MC,

$$PB = PB',$$

and consequently

$$AP + PB = AP + PB'.$$

The left side will be a minimum only when the right side is a minimum. But the shortest path between A and B' is the straight line joining them, so that the right, and consequently the left side, will be smallest when P is collinear with A and B'.

It remains merely to show that when APB' is a straight line, the angles made by AP and BP with the normal at P are equal. Study Figure 3.5. Since APB' is a straight line, MPA and $B'PC$ are vertical angles, so that

$$\angle APM = \angle B'PC.$$

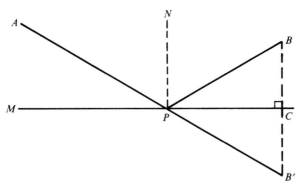

Figure 3.5

But

$$\angle BPC = \angle B'PC,$$

by symmetry (or by congruence of triangles PBC and $PB'C$, from two sides and their included angle, $PC = PC$, $\angle PCB = 90° = \angle PCB'$, $CB = CB'$). Hence

$$\angle APM = \angle BPC.$$

Therefore the complement of the former is equal to the complement of the latter, i.e.,

$$\angle APN = \angle BPN.$$

This completes the proof that E_2 is implied by H.

The critical reader may well ask, "How did Heron hit upon the idea of the auxiliary point B'?" But haven't we all seen swan and reflection floating double on a placid lake? The swan's image is the same size as the swan, but upside down. In terms of Figure 3.5, if MC represents the lake surface and CB the swan, then CB' represents the swan's image; in particular B' is the image of B. To an observing eye at A looking along AP, B appears to be on AP produced at B'. To see B in a reflecting surface is to see it as if it were at B' and there were no reflecting surface. The concept of mirror image enables us, in effect, to throw away the mirror and reduce the problem of a reflected ray's path to that of a nonreflected ray. By E_1 the path of light from A to B' (when no mirror intervenes) is the straight line AB'; the shortest path possible. We can but suppose that Heron had such considerations as these in mind when he pondered the problem.

3.3 Archimedes' Symmetry Proof

Symmetry pervades science. In the first chapter we saw how it pervades Archimedes' theory of equilibrium of the lever; in the final chapters we shall see further pervasions. Appropriately to conclude our examinations of the Law of Reflection we now consider Archimedes' proof. Not surprisingly in view of what we have already seen of Archimedes' "style", his proof is proof by symmetry.

Reconsider Figure 3.1. Suppose that l_1 represents an incident tube, a very narrow cylinder down which a beam of light is transmitted to the surface of a mirror, and l_2 represents a reflection tube, a very narrow cylinder up which the reflected beam travels. It is an experimental fact that light transmitted down l_2 is reflected up l_1; reflection tube becomes incident tube and incidence tube becomes reflection tube *without altering either angle α or β*.

Archimedes asks: Is the angle of incidence larger than the angle of reflection? If so, with l_1 the incidence tube, $\alpha > \beta$, and with l_2 the incidence tube, $\beta > \alpha$, so that α is both greater than and less than β. Similarly, the hypothesis that the angle of incidence is smaller than the angle of reflection is untenable. In short, by exploiting the symmetry of the situation, Archimedes drives us to the conclusion that $\alpha = \beta$.

3.4 Ptolemy and Refraction

The further development of optics leads us to the work of the great Alexandrian astronomer, Ptolemy, who flourished from 127 to 141 or 151 AD. Shortly after the time of Heron, deep interest in astronomy raised other questions concerning the nature of light. Ptolemy found from his observations of the stars that the propagation of light near the earth's surface is not precisely rectilinear, but slightly curved. On the analogy of the bent appearance of a straight stick partially immersed in water, he ascribed the curvature of light to passage through layers of air of different density.

Some textbook writers would have us believe that the Greeks were interested only in the things that they could see but not touch. On the contrary, a vast amount of experimental work was done in Alexandrian times. Ptolemy, to understand better the effect of change in density on the bending of light rays by the atmosphere, conducted experiments to measure the deflection of light rays in passing from air to water. See Figure 3.6.

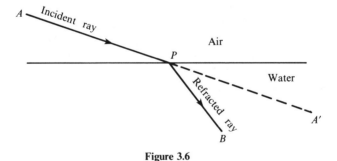

Figure 3.6

Upon penetrating the surface of the water, the incident ray does not continue along AP (produced), but at an angle to it. The deflected ray is said, to use the commonly accepted term, to be *refracted*. Possibly the reader is disposed to take $\angle BPA'$ as a measure of the refraction. Ptolemy did not do this. Refraction is sufficiently similar to reflection to merit analogous terminology. With both phenomena there is a ray incident to a surface, and therefore an angle of incidence. The only difference is that whereas with reflection the ray after incidence is deflected above the surface, with refraction it is deflected below it. Is it not therefore natural to measure angles for both phenomena with reference to the normal to the surface; to use the same definition of angle of incidence for both; and since a ray deflected upwards is measured against the upward normal, to measure a ray deflected downwards against the downward normal? Thus in Figure 3.7, α is said to be the *angle of incidence* and β the *angle of refraction*.

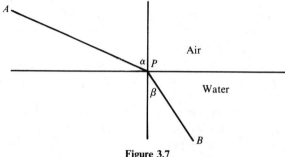

Figure 3.7

Ptolemy found that β depends upon α; a change in the angle of incidence results in a change in the angle of refraction. In mathematical terminology β is a function of α, say $\beta = f(\alpha)$. As a first step toward specification of $f(\alpha)$, Ptolemy made extensive tabulation of the ordered pairs, α with the corresponding β. Despite more and yet more experiments with extensive and yet more extensive tabulation, the law of ordering continued to elude him. Finally he had to give up.

3.5 Kepler and Refraction

More than a thousand years later the problem was tackled by Kepler (1571–1630), an astronomer, justly famous, who had a genius for finding the functional relation governing the most recalcitrant of ordered pairs. Allow us to illustrate his capacity.

Year after year he worked away, conjecturing and checking, until finally he hit upon hypotheses that fit his observational data. He showed that each planet describes an ellipse having the sun at one of its foci, and that the areas swept out by the radii drawn from the planet to the sun as it orbits the sun are proportional to the time taken to sweep them out. For each planet he knew the mean distance[†] r of its elliptical orbit from

---

[†]In this context, "mean distance" is to be interpreted as the arithmetic mean a of the shortest and longest distances of the planet from the sun (located at a focus). Thus

$$a = \tfrac{1}{2}(r_{\min} + r_{\max}) = \tfrac{1}{2}(2a) = \text{semi-major axis of the ellipse.}$$

MATHEMATICS IN OPTICS

the sun, and calculated the planetary year T, the time it takes to complete a full orbit. He tabulated T and r. He asked himself: "What is the functional relation between them?" He found the answer. Incredible man.

To get some measure of Kepler's achievement, the reader is asked to seek the relation of T to r in the Table 3.1.

Not obvious, eh? Alas, to find is to seek successfully. After hours or days of unsuccess we likely concede that such problems demand a Kepler. But these are neat and tidy figures, tailor-made for the occasion; devoid of messy decimals, our tabulation is designed to be more perspicuous than the observational data of Kepler's problem. His was difficult.

A hint. Our r column contains only perfect squares. Is the relation of T to r now obvious? No, the "obvious" conjecture is wrong; T is not also a perfect square. No, neither is T the sum of two perfect squares. T, it so happens, is a perfect cube. Advantageously we rewrite our tabulation in Table 3.2.

TABLE 3.1

T	r
287,496	484
1,601,613	1,521
2,146,689	1,849
4,251,528	2,916
4,721,632	3,136
7,414,875	4,225
9,261,000	4,900

TABLE 3.2

T	r
66^3	22^2
117^3	39^2
129^3	43^2
162^3	54^2
168^3	56^2
195^3	65^2
210^3	70^2

Is the relation between T and r now apparent to you? That depends upon your discernment. Perhaps you notice that, *neglecting exponents*, the first number of each ordered pair is three times the second, that

$$\frac{66}{22} = \frac{117}{39} = \frac{129}{43} = \cdots = 3.$$

Thus, for example,

$$66 = 3 \cdot 22,$$

so that the entry for T *including the exponent* is

$$66^3 = 3^3(22^3) = 3^3(22^2)(22).$$

What a pity the 22 is cubed instead of squared. Thinking wishfully, we write

$$(66^3)^2 = 3^6(22^2)^2(22)^2 = 729(22^2)^3.$$

It is left to the reader to show that

$$T^2 = 729 \cdot r^3$$

satisfies our tabulation.

Kepler's tabulation, though difficult, was governed by the same underlying proportionality. He found that

$$T^2 = kr^3,$$

where k is a constant. This is his famous third law that the square of the time of revolution of a planet about the sun is proportional to the cube of that planet's mean distance from the sun. Although our tabulation with nice whole numbers devoid of observational error inadequately illustrates his achievement, it does afford some hint of why Kepler's discovery cost him nearly a decade of incessant toil.

With equal enthusiasm Kepler turned to the refraction problem of specifying β in terms of α. Knowing his ability, we anticipate his success, yet even Kepler was unsuccessful.

3.6 Fermat: The Quickest Path Principle

Although the reader is understandably impatient to learn the correct formula, the development of science is not to be hurried. Solution of long standing problems is attendant upon the winds of fresh discovery; the new ideas of a lively intelligence, stimulated by the intellectual ferment of the day. The lively intelligence was Fermat's (Fermat 1601–1665); the intellectual ferment of his day was the question, "Does light have a velocity or is its propagation instantaneous?"

Possibly Galileo (1564–1642) was the first to tackle this question experimentally. At night on a mountain top he signalled with a lantern to a colleague on an adjacent mountain. His colleague, on seeing the light of Galileo's lantern, uncovered his own. Galileo tried to measure the interval between dispatch of his signal and receipt of his colleague's. As near as he could tell, light is instantaneous. To us the experiment is incredibly naive, but Galileo did not know that the time for, say two 10-mile light journeys, is of the order of one ten-thousandth of a second. He experimented to find out.

MATHEMATICS IN OPTICS

This live issue captured Fermat's attention. Suppose, he pondered, light is not instantaneously propagated, but has a velocity. Further suppose this velocity to be constant. What then? Time is distance divided by velocity, so the shortest path is the quickest. The supposition that light takes the shortest time has precisely the same consequences as Heron's principle that it takes the shortest path. But alternatively, suppose that the velocity of light, while constant in any given medium, is different for different media. In particular, suppose that light in water has a velocity different from light in air. What then? With travel in both air and water, the shortest path is not the quickest. A refracted ray, because refracted, does not take the shortest path: does it take the quickest? If so, in each medium, Heron's shortest path principle still holds, and perhaps refraction is also explicable.

Further thought gives this conjecture further plausibility. Heron's minimum principle that light takes the shortest path is more embracing and covers reflection as well as the hypothesis E_1 on page 77. Why not an even more embracing minimum principle that covers refraction as well as reflection and E_1?

Fermat's conjecture explains at least as much as Heron's. Does it explain more? What precisely are the implications for refraction of the minimum principle that light takes the quickest path possible? What is the quickest path?

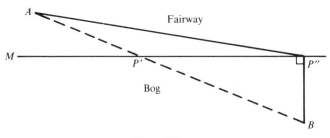

Figure 3.8

Consider the plight of a golfer, Colonel Bogey, who in driving from the fairway at A hooks his ball into the bog at B. See Figure 3.8. What is his quickest route to the ball? If bog is replaced by fairway so that Bogey's bog velocity v_2 is equal to his fairway velocity v_1, then obviously his quickest route is the shortest route $AP'B$. Next consider the extreme case in which the bog is so bad that it is all but impassable. In this case, while Bogey can gaily stride along the fairway at (say) 3 miles an hour, it is only with the fanaticism of a dedicated golfer that he can flounder through the bog chest deep at (say) three yards an hour. To avoid a mere three yards of bog Bogey could without loss of time afford to travel an extra three miles of fairway. In these arduous circumstances Bogey's quickest route is close to the path of minimum bog, i.e. that in which BP'' is perpendicular to MP'.

In brief, when $v_2 = v_1$, $AP'B$ is Bogey's quickest route; when v_2 is almost zero, his quickest route is almost $AP''B$. Next consider intermediate cases. Suppose that bog is replaced by bracken and gorse. Off the fairway the going is not so desperately slow as floundering through the bog, yet slower than fairway walking. The rougher the going is off the fairway, the more Bogey will have to deviate from the shortest route $AP'B$ to cut down the amount of rough, time-consuming terrain he must cross. Clearly, as v_2 decreases from v_1 to nearly zero, the point P of the quickest route APB moves from P' toward P''.

We suppose APB to be the quickest route from A on the fairway to B in the rough. See Figure 3.9.

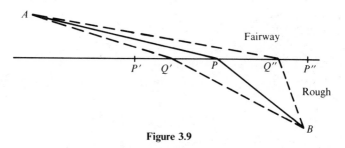

Figure 3.9

Consider any other path AQB crossing the dividing line between fairway and rough at a point Q. Since APB is the quickest path,

$$\text{travel time along } AQB \geq \text{travel time along } APB,$$

that is

$$\frac{AQ}{v_1} + \frac{BQ}{v_2} \geq \frac{AP}{v_1} + \frac{BP}{v_2}$$

or, equivalently

$$\frac{BQ - BP}{v_2} - \frac{AP - AQ}{v_1} \geq 0.$$

This excess travel time is a non-negative function of the crossing point Q. We denote it by $t(Q)$:

$$(3.1) \qquad t(Q) = \frac{BQ - BP}{v_2} - \frac{AP - AQ}{v_1} \geq 0.$$

Clearly it takes its minimum value $t(P) = 0$ when $Q = P$. This char-

MATHEMATICS IN OPTICS   87

acterization is sufficient to determine the quickest path from A to B. This is a very special case of minimum problems treated in differential calculus, and we shall treat it by quite elementary methods, with a little trigonometry.

We have written $t(Q)$ so that, when Q is to the left of P, both numerators in (3.1) are positive. Observe that while α is the fixed angle between the path AP and the perpendicular AM, the angle β depends on the crossing point Q and changes as Q moves. We indicate this dependence by writing β as a function $\beta(d)$ of the distance between Q and the fixed point P. See Figure 3.10.

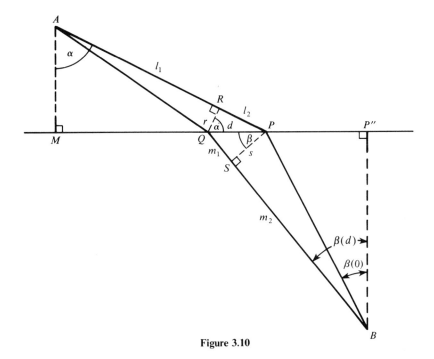

Figure 3.10

From Q we drop the perpendicular QR to AP; it splits segment AP into the two segments AR and RP whose lengths we denote by l_1 and l_2, respectively. Similarly, we drop the perpendicular PS from P to QB; it splits segment QB into segments QS and SB of lengths m_1 and m_2. We let d denote the distance of the variable point Q from the fixed crossing point P.

Considering right triangles AQR and PQR, we find that

$$r = d\cos\alpha, \qquad l_2 = d\sin\alpha,$$

and by the Pythagorean theorem applied to $\triangle AQR$,

$$AQ = \sqrt{l_1^2 + r^2} = l_1\sqrt{1 + \frac{d^2\cos^2\alpha}{l_1^2}}.$$

Since $AP = l_1 + l_2$, we find that

$$AP - AQ = l_1 + l_2 - AQ = l_1 + d\sin\alpha - l_1\sqrt{1 + (d^2\cos^2\alpha/l_1^2)}$$

or

(3.2) $$AP - AQ = d\sin\alpha - l_1\left[\sqrt{1 + (d^2\cos^2\alpha/l_1^2)} - 1\right].$$

We now consider the two right triangles QSP and BSP and find similarly that

$$s = d\cos\beta, \qquad m_1 = d\sin\beta$$

and

$$BP = \sqrt{m_2^2 + s^2} = m_2\sqrt{1 + (d^2\cos^2\beta/m_2^2)}.$$

Since $BQ = m_1 + m_2$,

(3.3) $$BQ - BP = d\sin\beta - m_2\left[\sqrt{1 + (d^2\cos^2\beta/m_2^2)} - 1\right].$$

Equations (3.2) and (3.3) involve terms of the form $\sqrt{1 + x} - 1$ with $x > 0$. Observe that

$$\left(1 + \frac{x}{2}\right)^2 = 1 + x + \frac{x^2}{4} \geq 1 + x;$$

taking positive square roots, we obtain

$$1 + \frac{x}{2} \geq \sqrt{1 + x}.$$

Thus

$$\sqrt{1 + x} - 1 \leq \frac{x}{2}.$$

To avoid inequality signs, we write this in the form

(3.4) $$\sqrt{1+x} - 1 = \zeta x, \qquad 0 \leq \zeta \leq \tfrac{1}{2}.$$

The exact value of the number ζ is not known, but we can use its upper and lower bounds in an elegant way, as we shall see.

Using first $d^2\cos^2\alpha/l_1^2$ then $d^2\cos^2\beta/m_2^2$ for x in (3.4), we rewrite formulas (3.2) and (3.3) in the form

(3.5) $$AP - AQ = d\sin\alpha - (d^2\cos^2\alpha/l_1^2)\zeta_1$$

and

(3.6) $$BQ - BP = d\sin\beta - (d^2\cos^2\beta/m_2^2)\zeta_2,$$

where ζ_1 and ζ_2 are real numbers between 0 and $\tfrac{1}{2}$.

We are now ready to use condition (3.1). We find for any shift d of the crossing point to the left

(3.7) $$\left[d\sin\beta - \frac{d^2\cos^2\beta}{m_2^2}\zeta_2\right]\frac{1}{v_2} - \left[d\sin\alpha - \frac{d^2\cos^2\alpha}{l_1^2}\zeta_1\right]\frac{1}{v_1} \geq 0.$$

Since d is a positive number, we may divide by it and obtain

(3.7′) $$\left[\sin\beta - \frac{d\cos^2\beta}{m_2^2}\zeta_2\right]\frac{1}{v_2} - \left[\sin\alpha - \frac{d\cos^2\alpha}{l_1^2}\zeta_1\right]\frac{1}{v_1} \geq 0.$$

Now, let $d \to 0$, and in the limit we find

(3.7″) $$\frac{\sin\beta}{v_2} - \frac{\sin\alpha}{v_1} \geq 0.$$

When $d \to 0$, $Q \to P$ and $\beta(d) \to \angle PBP''$. Thus the β appearing in the limit relation (3.7″) must be interpreted as the angle $\beta(0)$ between the fixed path BP and the line BP'' normal to the interface MP.

We have obtained the inequality (3.7″) by considering a shift of P to the left. Had we shifted P to the right, we would have lengthened the path AP and shortened the path PB. The same calculation could be carried out by interchanging the roles of A and B (i.e., looking at Figure 3.10 upside down) and would yield the inequality

(3.8) $$\frac{\sin\beta}{v_2} - \frac{\sin\alpha}{v_1} \leq 0.$$

We are thus led to the conclusion

$$(3.9) \qquad \frac{\sin \beta}{v_2} - \frac{\sin \alpha}{v_1} = 0,$$

and consequently to

$$(3.10) \qquad \frac{v_1}{v_2} = \frac{\sin \alpha}{\sin \beta}.$$

We have derived this law in the context of a game of golf. It is just as valid in the case of propagation of light. Replace the fairway, say, by the medium of air, the rough by the medium of water, the golfer Bogey by a ray of light, and you find in (3.10) Fermat's law for the refraction of a light ray in passing from one medium to another with a different velocity of light.

We note that when $v_1 = v_2$, as for example when both media are identical, the condition for the quickest path is also the condition for the shortest path, so that Heron's (shortest path) law of reflection can be regarded as a special case of Fermat's law of refraction.

Alternatively, (3.10) can be expressed in the form

$$(3.11) \qquad \sin \alpha = k \sin \beta, \qquad \text{where} \quad k = \frac{v_1}{v_2},$$

i.e. $\sin \alpha$ is directly proportional to $\sin \beta$, where the constant of proportionality k is the ratio of the velocity of light in the first medium to that in the second. Physicists call k the *index of refraction*.

We recall that Kepler, despite his ability and enthusiasm, failed to discover the law of refraction. He first conjectured the law to be of the form: α is proportional to β. Since $\sin \gamma \to \gamma$ as $\gamma \to 0$, Kepler's formula approximates (3.11) and is increasingly accurate as the angles are decreased. He subsequently conjectured

$$\alpha = a\beta + b \sec \beta,$$

where a and b are constants independent of α and β. This formula turns out to be preferable and (for example with a and b determined experimentally for refraction between air and water) to be quite accurate for $\alpha < 15°$. From (3.10) we get

$$\alpha = \arcsin\left(\frac{v_1}{v_2} \sin \beta\right).$$

That Kepler failed to conjecture such a complicated functional relationship between α and β occasions no surprise. What is surprising is that his second attempt was so successful.

Fermat's law was later rediscovered independently by both Snell and Descartes and was used by Descartes to explain the phenomenon of the rainbow. Fermat discovered his law (3.10), as we have just shown, by solving the problem of the quickest path. The method he invented, used above, was invented independently by Newton and by Leibniz and has been developed ultimately into what we now know as the calculus of variations. Few mathematicians, even among those of the first rank, could claim invention of the calculus. Fermat was one of the few.

3.7 Newton's Mechanistic Theory of Light

With Newton (1642–1727) science came of age. Understanding the heavens was within man's grasp. It was almost as if Newton with his three laws and few axioms could, as Jesus with his five loaves and two fishes, work miracles. He explained the ebb and flow of the waters of the deep and the passage of the planets in the firmament above. Nature lost her mystery; man his impotence. The solar system is a gigantic piece of clockwork, and Newton had discovered how it ticks. Newton's mechanics is the key to everything around the sun; must it not be the key to everything under the sun? Newton thought that the optics of Euclid, Heron, and Fermat could be explained mechanistically.

Newton begins at the beginning. The first thing to explain is the rectilinear propagation of light. His first law states that a body moving with uniform velocity in a straight line will continue to do so unless acted upon by external forces to change that motion. Are not Euclid's and Newton's first laws remarkably similar? With characteristic ingenuity Newton makes the former an immediate consequence of the latter by introducing the supposition that a ray of light consists of minute particles, or *corpuscles*. Because of this assumption Newton's theory of light is known as the corpuscular theory.

How does Newton account for Euclid's law of reflection? According to his corpuscular theory, an incident ray of light is reflected because of the collision of its constituent particles with those constituting the surface of the mirror. It is, of course, sufficient to consider the fate of a single incident particle, for all the others will behave in the same way under similar circumstances.

First let us consider a special case, that of an incident ray normal to the mirror. See Figure 3.11. What happens when a constituent particle of a ray travelling along NP reaches P? Its attempt to penetrate the surface MM' perpendicularly downwards is resisted *solely by forces acting perpendicularly upwards* (due to the constituent particles of the mirror surface in the neighborhood of P). Consequently the particle returns along the normal.

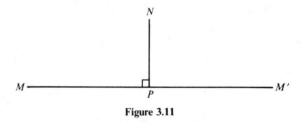

Figure 3.11

We now turn to the general case. It is assumed that v_1, the velocity of a light ray in air, is constant irrespective of its direction relative to the mirror. Thus the problem is the following. A particle travelling with velocity v_1 along AP at an angle α to the normal is reflected with velocity v_1 along PB which makes some angle β with the normal; see Figure 3.12.

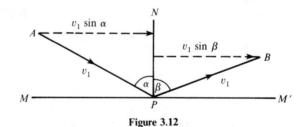

Figure 3.12

What is the relation between β and α? Despite the fact that the particle at P now attempts to penetrate MM' obliquely, Newton insists that the structure of this surface is such that the resistance to penetration is solely by forces acting perpendicularly upwards—just as in the special case considered above. What are the consequences of his insistence? Forces acting in the direction PN normal to the surface have no components in the horizontal direction MM', so that the horizontal component of velocity of the particle at P is unchanged by impact. Therefore the velocity component parallel to MM' of the particle, when part of the reflected ray, is just the same as when part of the incident ray. *Motion parallel to the surface remains unchanged.* Equating component velocities parallel to MM', we have

$$v_1 \sin \beta = v_1 \sin \alpha,$$

and hence $\beta = \alpha$.

MATHEMATICS IN OPTICS

Next, refraction. What is the difference between reflection and refraction? Whereas in the latter the incident ray is successful in penetrating the surface, in the former it is not. Newton treats these phenomena similarly. No matter whether or not penetration is successful, Newton continues to insist that the only forces opposing penetration act perpendicularly to the surface. Consequently, for refraction as for reflection, motion parallel to the surface remains invariant. And while the refracted ray differs from the reflected ray by being propagated in water instead of air, so that its component velocity parallel to MM' is $v_2 \sin \beta$ instead of $v_1 \sin \beta$, the incident ray is the same in both cases. See Figure 3.13. Therefore, equating component velocities parallel to MM',

$$v_2 \sin \beta = v_1 \sin \alpha,$$

giving

(3.12) $$\sin \beta = K_2 \sin \alpha, \quad \text{where} \quad K_2 = \frac{v_1}{v_2}.$$

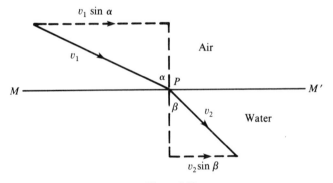

Figure 3.13

3.8 Fermat Versus Newton: Experimentum Crucis

Thus Newton, like Fermat, concludes that $\sin \beta$ is directly proportional to $\sin \alpha$. However, comparison of (3.11) with (3.12) also shows that although their formulae have the same form, Newton's constant of proportionality is the reciprocal of Fermat's. And it is an experimental fact that in going from air to water, the refracted ray is bent towards the normal, i.e. $\beta < \alpha$, so that $\sin \beta < \sin \alpha$. Consequently a formula of the form

$$\sin \beta = K \sin \alpha$$

cannot be correct unless the constant of proportionality K is less than unity. If (3.11) is correct, then $v_2/v_1 < 1$; if (3.12) is correct, then $v_1/v_2 < 1$. Whereas Fermat's formulae cannot be correct unless the velocity of light in water is less than in air, Newton's cannot be correct unless the precise opposite is the case.

Newton had no difficulty in finding an argument to vindicate his own theory. A particle of light when in air is travelling in a homogeneous medium, so that the forces acting upon it are constant; there being no acceleration, the net force must be zero. Similarly in water. But when a particle passes from one medium to another there is a change from one homogeneity to another, so that the forces acting upon it momentarily are not constant. Water has greater density than air; its particles are more tightly packed. When a particle reaches the neighborhood of the interface, ahead of it is an accumulation of matter, behind it a sparsity. Consequently, since the more the mass the greater the attraction, the particle has a momentary acceleration and speeds up from v_1 to v_2. Having passed through the interface, once again there is no net force and the particle continues with constant velocity v_2.

It was a good argument as long as it lasted; it lasted rather more than a century. And then technological advances made it possible to show experimentally that light is slower in water than in air. This was the *experimentum crucis* which showed definitely that Newton's theory could not be the explanation of the refraction phenomenon and that the Fermat principle was the correct description.

Newton's theory of light which wanted to explain the phenomena of optics by the motion of particles obeying Newtonian mechanics is called the corpuscular theory of light. The Dutch scientist Christian Huygens developed a rival theory in which the motion of light is explained as a wave motion of an all-pervasive fluid which he called the ether. The mathematical development of this so-called wave theory led indeed to the consequence that light rays obey the Fermat principle. This fact was one of the main arguments for adopting the wave theory and for rejecting the corpuscular theory.

Additional physical evidence followed soon. In 1802 Thomas Young succeeded in showing that two interfering light beams can cancel each other out and create total absence of light. This effect is easily understood by the wave theory and had been indeed predicted; it is not explainable by the corpuscular theory.

But ideas in science never die. In the beginning of the 20th century two new great theories were born; the quantum theory of Planck and the theory of relativity of Einstein. Relativity theory showed convincingly that the ether concept was meaningless, since in principle no experiment could be devised to confirm it. Quantum theory postulated that light consists of beams of particles, the so-called photons, which move with the speed of light. So we were back at the old Newtonian theory. It was, of course, necessary to reformulate classical mechanics to a new quantum mechanics, the so-called wave mechanics, to bring about the synthesis of the theories of Newton and Huygens. This is a good example of the continuous improvement and adjustment to new facts in science. We have a successive approximation to truth.

Let us finally remark that Fermat enunciated his quickest path principle a quarter of a century before it was known experimentally that the propagation of light is not instantaneous. Here, as is often the case in physics, the beauty of the mathematical description led to certain hypotheses which came before the experimental verification.

3.9 To Recapitulate

We have traces the development of elementary optics over the centuries up to the formulation of Fermat's and Newton's theories. Both explain rectilinear propagation of light; both account for the law of reflection; both give the same kind of formula for refraction. Yet they are rivals, for with regard to refraction they differ in detail. Here is a situation typical of the history of science, a conflict of theory only to be resolved by determination of fact. That is the role of crucial experiment.

But when a consequence of a theory is in question, the basis of the theory is also in question. In rejecting Newton's consequences for refraction as contrary to fact, we must reject the basis of these consequences—the corpuscular nature of light. That Fermat's theory could explain all the facts vindicated his quickest path principle. Later this developed into the wave theory of light—that light is constituted not of corpuscles, but of waves. Although space does not permit consideration of further developments, we hope to have shown you something well worth showing of the role of mathematics in the evolution of science. Mathematics sharpens our perception of logical consequences and focuses our attention on appropriate experimentation; an aid to vision, it is the eyeglass of the mind.

3.10 The Role of Science in Mathematics

From the role of mathematics in science, we turn to the role of science in mathematics; for despite an abundance of material, the way in which science leads to mathematical theorems is little known. Convenient to our purpose is the problem of how to construct a tangent to an ellipse.

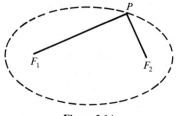

Figure 3.14

If two pegs F_1, F_2 are hammered into the ground and a cord tied to both of them is kept taut by a stick P, the movement of P under this restraint marks out an elliptical flower bed. See Figure 3.14. This method of construction exhibits one definition of the *ellipse*: The set of all points P with the property that the sum of the distances from P to two fixed points F_1, F_2 (called the foci) is constant, is said to be an ellipse. Traditionally the constant sum is taken to be $2a$, giving the equation of the ellipse as

$$F_1P + PF_2 = 2a.$$

The circle is a special case of the ellipse. When F_1, F_2 become coincident

$$F_1P + PF_2 = 2 \cdot F_1P = 2a,$$

so that F_1 (and F_2) become the center of a circle of radius a. This suggests that properties of the circle will be limiting cases of properties of the ellipse. What light does this suggestion throw on the problem of constructing a tangent at P to the ellipse?

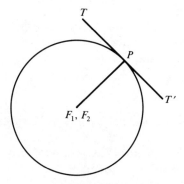

Figure 3.15

MATHEMATICS IN OPTICS

The tangent at P to the circle with center F_1 is perpendicular to the radius F_1P; see Figure 3.15. How do we go from this limiting case to the general case? Equally well we could say that the tangent is perpendicular to F_2P, or that it is perpendicular to both F_1P and F_2P. But obviously the tangent can be perpendicular only to one of these lines when they are no longer coincident. Which one? Surely they have equal claims. What is an acceptable compromise? That F_1P, F_2P are equally inclined to the tangent, see Figure 3.16, i.e., that $\gamma = \delta$. This conjecture has merit, for it is consistent with the limiting cases when F_1, F_2 become coincident and $\gamma = \delta = 90°$. But to suppose that $\gamma = \delta$ is equivalent to supposing their complements to be equal, i.e., that $\alpha = \beta$; see Figure 3.17. Are not the principal ingredients of this figure familiar? Could it not be interpreted as illustrating the law of reflection? We now use science to do mathematics.

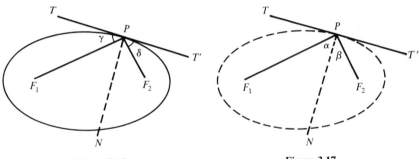

Figure 3.16   Figure 3.17

It is intuitively clear that a tangent to an ellipse has only one point, the point of tangency P, in common with the ellipse. All other points of the tangent line lie outside the ellipse. We suppose a ray of light F_1P to be reflected at P from a mirror TT' tangent to the ellipse at P. If the reflected ray does in fact pass through F_2, then TT' is perpendicular at P to the bisector of $\angle F_1PF_2$, and we have solved our problem.

Does the reflected ray pass through F_2? We recall that reflection is a consequence of the shortest path principle. Thus if Q is the point on the mirror from which an incident ray F_1Q is reflected through F_2, then F_1QF_2 must be the shortest path possible (via the mirror) from F_1 to F_2; see Figure 3.18. It remains to show that the shortest path is such that $Q = P$, the point at which TT' is tangent to the ellipse. It is evident that any point Q (on TT') other than P must lie outside the ellipse; therefore suppose F_1Q to cut the ellipse at R. Since RF_2 is the shortest path from R to F_2,

$$RQ + QF_2 > RF_2.$$

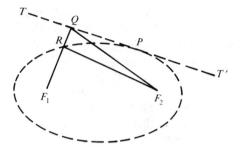

Figure 3.18

Consequently, adding F_1R to both sides of the inequality, we get

$$(F_1R + RQ) + QF_2 > F_1R + RF_2,$$

i.e.,

$$F_1Q + QF_2 > F_1R + RF_2.$$

But R is on the ellipse, so that by definition

$$F_1R + RF_2 = 2a.$$

Therefore,

$$F_1Q + QF_2 > 2a,$$

whereas, P being on the ellipse,

$$F_1P + PF_2 = 2a.$$

Since light takes the shortest path, it follows that Q must coincide with P. That is, a ray of light from one focus, incident to a mirror (tangent to the ellipse) at its point of contact, is reflected through the other. This completes the proof of our conjectured construction of a tangent to the ellipse.

Fermat was the man who first raised and substantially answered the wider question of how to find tangents to plane curves in general. To solve this problem for any curve which is the graph of a polynomial he invented the differential calculus. Yet it is refreshing with the present density of calculus textbooks to find that a construction for tangents to the ellipse can be established without resort to differentiation. The solution by optics, given above, was the earliest.

3.11 Some Practical Applications of Conics

This reflection property of an ellipse has several practical applications. The key to these applications is that for a silvered ellipse the immediate (elliptical) neighborhood of P will reflect light as if it were the surface at P of the mirror tangent to the ellipse at that point. Consequently, no matter what its direction, a ray passing through one focus will be reflected at the ellipse through the other. The heat of a fire at F_1, although radiated in all directions, will be reconcentrated at F_2. If no radiation is dissipated *en route* and none lost in contact with the ellipse's silvered surface, F_2 is as hot as F_1. A reflecting ellipse with a fire at one focal point has a fire at both. (*Focus* is the Latin for fireplace or hearth.) Similarly for an auditorium with an ellipsoidal cupola, F_1, F_2 are known as the whispering points. Since sound is reflected in the same way as light, a dispersed and therefore weakened whisper from F_1 will be inaudible in all other parts of the room except at F_2 where the whisper is reconcentrated. See Figure 3.19.

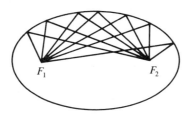

Figure 3.19

It is often instructive to go to the limit. We found it profitable to consider the limiting or degenerate case of the ellipse where F_1 and F_2 become coincident; we now go to the other extreme and suppose them to be as far apart as possible. With F_1 fixed, the farther F_2 is moved from it, the more elongated the ellipse and the more nearly parallel PF_2 to the axis, the line through the foci. See Figure 3.20. Finally, with F_2 at infinity, the ellipse has degenerated into what is known as a *parabola* and PF_2 has become parallel to the axis; see Figure 3.21. Thus, given a point source of light at F_1, the reflection from a silvered parabola MVM' is a beam parallel to the axis VF_1 of the parabola. Rotation of the parabolic mirror about its axis generates what is known as a paraboloid of revolution.

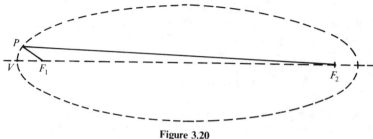

Figure 3.20

This, of course, reflects a solid beam of light from a point source at F_1 parallel to its axis, and is exemplified by an automobile headlight. Conversely, since the rays radiated from a distant source are almost parallel, they accumulate in the immediate neighborhood of F_1. A paraboloidal reflector could with equal justice be termed a paraboloidal accumulator. Radio rays, individually weak, can be collectively magnified into a strong signal. As well as being essential to radar listening devices, the paraboloidal reflector is the basis of the radio telescope.

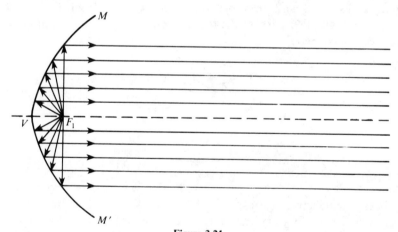

Figure 3.21

3.12 Conical Ingenuity; the Reflecting Telescope

The reflecting telescope is in many ways superior to the refracting telescope, because it uses a single reflecting surface rather than a system of several refracting lenses. Essentially it consists of a paraboloidal mirror at the bottom of a cylindrical tube. The mirror reflects onto one image point, the focus F, the nearly parallel rays that enter the tube. See Figure 3.22. So far, so good; but how can we see the focus from outside?

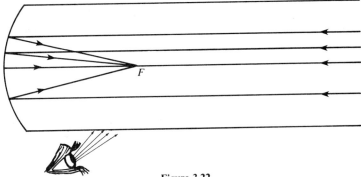

Figure 3.22

Newton solved the problem in the following way. He placed a small plane reflecting surface M close to the focus F to reflect the image to the side of the telescope; then to observe the image he made a hole in the side. See Figure 3.23. Mindful of the properties of mirrors, note that the plane of the mirror bisects FF' perpendicularly; that is, F' is as far in front of M as F is behind it.

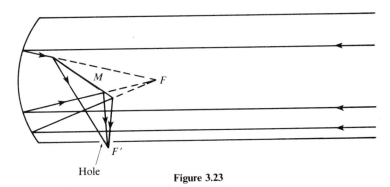

Figure 3.23

The French astronomer Cassegrain proposed an ingenious alternative arrangement. To appreciate his ingenuity we now need to turn to the geometrical properties of the last type of conic section, the hyperbola.

To define this curve, as in the case of the ellipse, we again introduce two fixed points F_1 and F_2, called its foci. The hyperbola is the set of all points in the plane satisfying the condition $|PF_1 - PF_2| = 2a$ for a fixed value of the parameter a. It consists of two distinct branches: a right branch for which $PF_2 - PF_1 = 2a$, and a left branch for which

$PF_1 - PF_2 = 2a$; see Figure 3.24. These separate the plane into three disjoint regions; one containing F_1, another containing F_2, and a third which contains neither F_1 nor F_2.

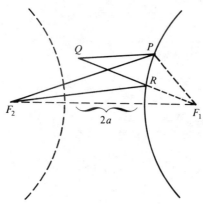

Figure 3.24

To work towards Cassegrain's idea we ask: If the right branch is a reflecting mirror and the left branch transparent, and if Q is a point in the third region, at what point R ("R" for point of Reflection) on the right branch would a ray of light from Q be reflected to pass through F_2? The answer to this question is dictated by our minimum principle in optics: The required point R is the point that minimizes the length of the path from Q to F_2 via a point on the right branch of the hyperbola.

We claim that R is the point where the segment joining Q and F_1 intersects the hyperbola. To prove this claim we let P be any other point on the right branch of the hyperbola and compare the lengths of the paths QPF_2 and QRF_2; see Figure 3.24.

If $P \neq R$, we have

$$QP + PF_1 > QF_1,$$

and since Q, R and F_1 are collinear, $QF_1 = QR + RF_1$. Therefore

(3.13) $$QP + PF_1 > QR + RF_1.$$

Since P and R are both on the right branch of the hyperbola, $PF_2 - PF_1 = 2a$ and $RF_2 - RF_1 = 2a$. Therefore

(3.14) $$PF_2 - PF_1 = RF_2 - RF_1.$$

Adding (3.13) and (3.14), we get

$$QP + PF_2 > QR + RF_2.$$

Since P is any point other than R on the right branch of the hyperbola, the last inequality proves our claim. This result also shows that the foci of the hyperbola have properties somewhat analogous to those of the ellipse.

We are now ready for Cassegrain's application. He placed, close to the focus F_1 of the reflecting mirror (which is a paraboloid of revolution), a small mirror which is part of one branch of a hyperboloid of revolution, obtained by rotating a hyperbola with foci F_1, F_2 about the axis $F_1 F_2$. The rays reflected from the paraboloidal mirror toward its focus F_1 are, because F_1 is also the focus of the hyperboloidal mirror, reflected from it toward the other hyperboloidal focus F_2. A small hole at the center of the paraboloidal mirror allows the rays to reach the observer's eye at F_2. See Figure 3.25.

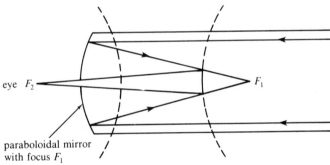

Figure 3.25

It remains only to remark that by increasing the parameter a (see Figure 3.24) for a given, fixed pair of foci F_1, F_2, the right, reflecting branch of the hyperbola can be brought closer to F_1 (and the left, transparent branch closer to F_2), thereby allowing much greater flexibility in construction than with a Newtonian reflector.

Thus, we have brilliant application of a little geometry to optics—with astronomical success. The 200-inch Mount Palomar telescope, which employs several Cassegrain mountings to give different focal lengths and magnifications for different purposes, can reach out to a distance of 1 billion (i.e. 10^9) light-years. A light year is the distance light travels in 1 year, namely 5.87 trillion (5.87×10^{12}) miles.

CHAPTER FOUR

Mathematics with Matrices—Transformations

Our main purpose in this chapter and the next is to show what matrices are good for; but before we can use this powerful tool we must understand how it works. We shall therefore begin by constructing it, starting with familiar pieces and showing, along the way, new efficient ways of looking at and talking about geometry, trigonometry, and their merger with algebra. In Chapter 5 we aim to show how this facility, used with bold imagination, devastates comfortable, commonplace conceptions of our physical world.

4.1 Why Use Matrices?

Have you ever tried using a lump of rock to drive a six-inch nail into a four-inch beam? It is easier with a hammer. Easier, because the hammer is designed expressly for the job, designed to have good balance, to handle well, to effect a neater job with less effort. Its design, deceptively simple, is the result of much thought about questions of rigidity, distribution of weight and center of percussion. Hard thinking goes into its design; hard work is eased by its use.

Matrices, too, are deceptively simple. Clever people have given much thought to devising a notation that handles well and a technique that does a tidier, more effortless job. Yes, matrices take the slog out of nailing equations. And, as with driving nails, there is no need to take anyone's word for it; experience is conclusive.

4.2 Plane Analytic Geometry and Vector Addition

Descartes invented analytic geometry in order to express geometric relations in terms of numbers, thus making geometric problems accessible

MATRICES—TRANSFORMATIONS

to algebraic methods. He chose a point O (origin) and a pair of perpendicular number lines (axes) intersecting at O. To any point P in the plane, he assigned a pair of numbers (x, y) called the coordinates of P, in the well-known fashion (see Figure 4.1). Let r be the length of the line segment OP, and let θ be the angle OP makes with the positive X-axis. By Pythagoras' theorem, $r = \sqrt{x^2 + y^2}$; moreover, the coordinates satisfy

(4.1) $\qquad x = r\cos\theta, \qquad y = r\sin\theta.$

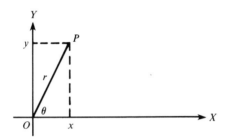

Figure 4.1

The algebraization of geometry was later aided by the invention of the notion of vector. There are many ways of introducing this concept; here we shall identify vectors with *displacements* of the plane. A displacement is defined as a shifting of the plane in any direction and by any amount. In a displacement, each point P is mapped onto some other point P' called the *image* of P. For any points P and Q, the segments PP' and QQ' are parallel and congruent, so that the four points P, P', Q, Q' form a parallelogram; see Figure 4.2. Therefore if we know the image P' of a point P under a displacement, we can determine the image Q' of any other point Q.

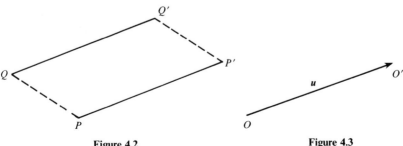

Figure 4.2   **Figure 4.3**

When discussing relations between various displacements it is convenient to describe all of them by specifying the images of a common reference point, namely the origin O. The directed line segment from O to its image O' is called the *vector* representing the displacement; this line segment is usually drawn with an arrow as in Figure 4.3, and denoted by a single letter, say *u*. The length of OO' is called the *length* of the vector *u* and is denoted by $|u|$.

Suppose two displacements are given; if we perform one and then the other, we obtain a new displacement called the *sum* of the first two. If *u* and *v* are the vectors representing the given displacements, the vector representing their sum is the diagonal of the parallelogram with sides *u* and *v*, and is denoted by $u + v$; see Figure 4.4.

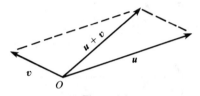

Figure 4.4

We claim that vector addition, as just defined, has all the usual properties of addition. The commutative law,

(4.2) $$u + v = v + u,$$

follows immediately from the parallelogram property of displacements.

The zero displacements is no displacement at all. It is represented by the zero vector, which consists of the single point O and is denoted by *o*. It behaves as it should:

(4.3) $$u + o = u.$$

For every displacement there is an *inverse* displacement which moves each point back where it came from. If the vector *u* represents the given displacement, its inverse is represented by the vector $-u$. Clearly

(4.4) $$u + (-u) = o,$$

as illustrated by Figure 4.5, which also shows $v + (-u) = v - u$, a vector parallel and congruent to the segment connecting the endpoint of *u* to the endpoint of *v*.

MATRICES—TRANSFORMATIONS

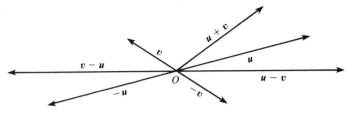

Figure 4.5

For a positive integer k we define $k\mathbf{u}$ by repeated addition: $k\mathbf{u} = \mathbf{u} + \mathbf{u} + \cdots + \mathbf{u}$. More generally for any positive number k we define $k\mathbf{u}$ as the vector pointing in the same direction as $\mathbf{u}$ and k times as long as $\mathbf{u}$. For k negative we define $k\mathbf{u}$ to be the vector $|k|(-\mathbf{u})$. It is easy to verify that the product of a vector by a number, as just defined, has the usual properties:

(4.5) $\qquad (k+l)\mathbf{u} = k\mathbf{u} + l\mathbf{u}, \qquad (kl)\mathbf{u} = k(l\mathbf{u})$

and

(4.6) $\qquad k(\mathbf{u} + \mathbf{v}) = k\mathbf{u} + k\mathbf{v}.$

4.3 The Dot Product

To complete the algebra of vectors, we define a multiplication of two vectors which may look a bit strange at first, but which turns out to have nice geometric consequences and furnishes a powerful tool for later use. First we define this product in the case where the two factors are equal, as

(4.7) $\qquad \mathbf{u} \cdot \mathbf{u} = |\mathbf{u}|^2, \qquad$ where $|\mathbf{u}|$ is the length of $\mathbf{u}$.

So, in this definition, *the product of a vector by itself is a number*. To define the product of two distinct vectors, we assume two usual algebraic properties of a product, the commutative and the distributive laws:

(4.8) $\qquad \mathbf{u} \cdot \mathbf{v} = \mathbf{v} \cdot \mathbf{u}, \qquad \mathbf{u} \cdot (\mathbf{v} + \mathbf{w}) = \mathbf{u} \cdot \mathbf{v} + \mathbf{u} \cdot \mathbf{w}.$

We use these to compute

$$(\mathbf{u} + \mathbf{v}) \cdot (\mathbf{u} + \mathbf{v}) = \mathbf{u} \cdot \mathbf{u} + \mathbf{u} \cdot \mathbf{v} + \mathbf{v} \cdot \mathbf{u} + \mathbf{v} \cdot \mathbf{v}$$

or

$$|u + v|^2 = |u|^2 + 2u \cdot v + |v|^2,$$

whence

(4.9) $$2u \cdot v = |u + v|^2 - |u|^2 - |v|^2.$$

This strange-looking expression gives the product $u \cdot v$ as a number, namely half the right member of eq. (4.9). This product of u and v, because of the notation used, is termed their *dot product*. To understand its geometric significance, we turn to the connection between two brilliant inventions: Descartes' coordinate geometry and the new vector algebra.

4.4 To Relate Coordinate Geometry and Vector Algebra

Having chosen an origin O for vectors, we now use the same point O as the origin of a coordinate system, choosing any pair of perpendicular lines through O as axes. Any vector v has initial point $(0, 0)$ and endpoint (x, y); x and y are called the *components* of v with respect to the chosen coordinate system, and this relation is written[†]

$$v = \begin{pmatrix} x \\ y \end{pmatrix}.$$

For example, let e and f be unit vectors (vectors of length 1) pointing in the direction of the positive X- and Y-axis, respectively. According to the above definition, we may represent them in terms of their components as

(4.10) $$e = \begin{pmatrix} 1 \\ 0 \end{pmatrix}, \qquad f = \begin{pmatrix} 0 \\ 1 \end{pmatrix}.$$

Let

$$v_1 = \begin{pmatrix} x_1 \\ y_1 \end{pmatrix} \quad \text{and} \quad v_2 = \begin{pmatrix} x_2 \\ y_2 \end{pmatrix}.$$

By the parallelogram law of vector addition, we deduce that

$$v_1 + v_2 = \begin{pmatrix} x_1 + x_2 \\ y_1 + y_2 \end{pmatrix};$$

[†] For some purposes row vectors $v = (x, y)$ are convenient; we prefer column vectors $v = \begin{pmatrix} x \\ y \end{pmatrix}$ for later use in matrix multiplication.

see Figure 4.6. In words: the components of the sum of two vectors are the sums of the components of the two vectors. More concisely: *To add two vectors, add their components.*

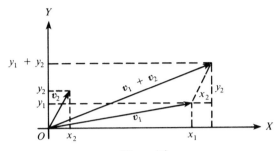

Figure 4.6

Figure 4.7 illustrates that for $v = \begin{pmatrix} x \\ y \end{pmatrix}$ and any real number k,

$$kv = \begin{pmatrix} kx \\ ky \end{pmatrix}.$$

In words: *To multiply a vector by a number, multiply each of its components by that number.*

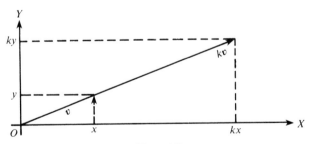

Figure 4.7

From these rules of addition and multiplication and from the definition of e and f in (4.10), we deduce that if $v = \begin{pmatrix} x \\ y \end{pmatrix}$, then

(4.11) $\quad xe + yf = x\begin{pmatrix} 1 \\ 0 \end{pmatrix} + y\begin{pmatrix} 0 \\ 1 \end{pmatrix} = \begin{pmatrix} x \\ 0 \end{pmatrix} + \begin{pmatrix} 0 \\ y \end{pmatrix} = \begin{pmatrix} x \\ y \end{pmatrix} = v.$

We know that the length $|v|$ of a vector $v = \begin{pmatrix} x \\ y \end{pmatrix}$ satisfies the relation

$$|v|^2 = x^2 + y^2.$$

Using this relation and our definition (4.9) of the dot product of two vectors

$$v_1 = \begin{pmatrix} x_1 \\ y_1 \end{pmatrix} \quad \text{and} \quad v_2 = \begin{pmatrix} x_2 \\ y_2 \end{pmatrix},$$

we get

$$2v_1 \cdot v_2 = |v_1 + v_2|^2 - |v_1|^2 - |v_2|^2$$
$$= (x_1 + x_2)^2 + (y_1 + y_2)^2 - (x_1^2 + y_1^2) - (x_2^2 + y_2^2)$$
$$= 2x_1 x_2 + 2y_1 y_2$$

so that

(4.12) $$v_1 \cdot v_2 = x_1 x_2 + y_1 y_2.$$

In words: *To obtain the dot product of two vectors, multiply the corresponding components and add the products.*

Note that it follows from (4.12) that the dot product is commutative and distributive. This is not surprising, since it was designed so as to satisfy these laws; see equations (4.8).

4.5 The Law of Cosines Revisited

Formula (4.12) holds with respect to any coordinate system. This observation enables us to get a coordinate-free formula for a dot product as follows: We choose the X-axis so that the endpoint of v_1 lies on the positive X-axis; see Figure 4.8. Then

$$v_1 = \begin{pmatrix} |v_1| \\ 0 \end{pmatrix}.$$

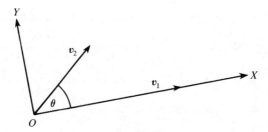

Figure 4.8

Denote by θ the angle between v_1 and v_2. Using relations (4.1) for Cartesian coordinates, we see that

$$v_2 = \begin{pmatrix} |v_2|\cos\theta \\ |v_2|\sin\theta \end{pmatrix}.$$

Using (4.12) and these component representations of v_1 and v_2, we find

(4.13) $\qquad v_1 \cdot v_2 = |v_1|\,|v_2|\cos\theta.$

In words: *The dot product of two vectors is equal to the product of their lengths times the cosine of the angle between them.* If the components of v_1, v_2 are known, (4.12) and (4.13) may be used to find the cosine of the angle between v_1 and v_2.

Consider a triangle formed by O and the endpoints of two arbitrary vectors u and v. The lengths of its sides are $|u|, |v|$ and, see Figure 4.5, $|u - v|$. Calculating the square of the length $|u - v|$ by means of the dot product, we get

$$|u - v|^2 = (u - v) \cdot (u - v) = |u|^2 + |v|^2 - 2u \cdot v.$$

Setting the lengths of the sides of the triangle equal to a, b, c, respectively, and using (4.13), we have proved

$$c^2 = a^2 + b^3 - 2ab\cos\theta,$$

which the reader probably recognizes as the *law of cosines*.

Another consequence of (4.13) is that if the dot product of two vectors is 0, then either $\cos\theta = 0$ (the angle θ between them is 90°), or one of the vectors is the zero vector. Two vectors whose dot product vanishes are called *orthogonal* (from the Greek word for perpendicular). It follows from this definition that the zero vector is orthogonal to every vector v since $o \cdot v = 0$ for every v.

Formula (4.13) also furnishes some qualitative information; e.g. if two vectors have a positive dot product, then $\cos\theta > 0$, i.e. the angle between them is acute. Similarly, if their dot product is negative the angle between them is obtuse.

We can also deduce from (4.13) that the dot product of two vectors u and v never exceeds the product of their lengths. For since $\cos\theta \le 1$, we have

$$u \cdot v \le |u|\,|v|.$$

4.6 Linear Transformations of the Plane

We turn now to studying mappings of the plane into itself. A *map M* (also called *mapping* or *transformation*) of the plane assigns to each point P of the plane a point Q, called the *image* of P under the map M; this assignment is denoted by

$$M(P) = Q.$$

After choosing an origin, we can identify any point P with the endpoint of a vector OP which we designate by u, and we identify its image point $M(P) = Q$ with the endpoint of the vector OQ which we denote by v; then we can think of a map as assigning to each vector u an image vector v:

$$M(u) = v.$$

We shall consider a particularly important class of maps called *linear maps* and characterized by these two properties:

(4.14) $$M(u_1 + u_2) = M(u_1) + M(u_2),$$

(4.15) $$M(ku) = kM(u),$$

called the additive property and the homogeneity property, respectively. In words: In a linear transformation, the image of a sum is the sum of the images, and the image of a multiple of a vector is that multiple of the image.[†] A linear transformation maps the origin into itself. This can be seen for example by writing $o = u - u$ so that

$$M(o) = M(u - u) = M(u) - M(u) = o.$$

Note that property (4.14) is a distributive property, while (4.15) says that multiplying by k commutes with applying M. Thus the action of M on vectors has some of the attributes of multiplication; for this reason it is customary to denote the action of a linear map M as multiplication, to drop the parentheses and simply write

$$M(u) = Mu.$$

[†] The term "linear" is sometimes misused to denote functions of the form $f(x) = mx + b$, where m, b are constants; however, if $b \neq 0$, $f(x)$ fails to have properties (4.14), (4.15). The analogue in the plane would be transformations of the form $T(u) = M(u) + b$, i.e. a linear map followed by a displacement. Such transformations are called *affine maps*.

Let $v = \begin{pmatrix} x \\ y \end{pmatrix}$ be any vector; according to (4.11) we can express v by

$$v = xe + yf.$$

Let M be any linear map; using properties (4.14), (4.15) we see that

(4.16) $$Mv = M(xe + yf) = xMe + yMf.$$

This shows that if we know how M acts on e and f, then we can determine how M acts on any vector v. Let us name the components of the images of e and f under M:

$$Me = \begin{pmatrix} a \\ c \end{pmatrix}, \quad Mf = \begin{pmatrix} b \\ d \end{pmatrix}.$$

Then by (4.16), we may write

(4.17) $$M\begin{pmatrix} x \\ y \end{pmatrix} = x\begin{pmatrix} a \\ c \end{pmatrix} + y\begin{pmatrix} b \\ d \end{pmatrix} = \begin{pmatrix} ax + by \\ cx + dy \end{pmatrix}.$$

The action of M is completely determined by the four numbers a, b, c, d, and it is customary to write M as a *matrix*

$$M = \begin{pmatrix} a & b \\ c & d \end{pmatrix}$$

whose action on vectors is defined by (4.17) and called multiplication:

(4.17′) $$M\begin{pmatrix} x \\ y \end{pmatrix} = \begin{pmatrix} a & b \\ c & d \end{pmatrix}\begin{pmatrix} x \\ y \end{pmatrix} = \begin{pmatrix} ax + by \\ cx + dy \end{pmatrix}.$$

Two matrices are *equal* if they define the same map; that is, if each element of one is equal to the corresponding element of the other.

Note that Mv, the product of a matrix M and a vector v, is a vector whose first component is the dot product of the first row of M with v and whose second component is the dot product of the second row of M with v.

We have proved that to every linear map M there corresponds a matrix

$$M = \begin{pmatrix} a & b \\ c & d \end{pmatrix}$$

with the action of the map on any vector $\begin{pmatrix} x \\ y \end{pmatrix}$ given by equation (4.17′).

4.7 Rotations

A natural and useful example of a map is rotation of the plane around the origin through some angle ϕ; we call this map $R[\phi]$. See Fig. 4.9. We claim that $R[\phi]$ is linear, i.e. that it has properties (4.14) and (4.15). To see that $R[\phi]$ is additive, consider two arbitrary vectors u, v and their sum $u + v$. When u and v are rotated about the origin by an angle ϕ, the entire parallelogram with sides u, v and diagonal $u + v$ is rotated by ϕ; the rotated diagonal satisfies

$$R[\phi](u + v) = R[\phi]u + R[\phi]v.$$

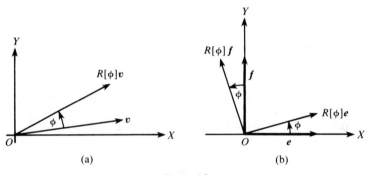

Figure 4.9

To see that $R[\phi]$ is a homogeneous map, we observe that rotating a multiple ku of u yields the same vector as multiplying the rotated vector $R[\phi]u$ by k:

$$R[\phi]ku = kR[\phi]u.$$

Having established the linearity of the rotation $R[\phi]$, we can now represent it as a matrix merely by observing the action of $R[\phi]$ on the unit vectors e and f as can be observed from Figure 4.9(b) and a simple calculation:

$$R[\phi]e = R[\phi]\begin{pmatrix}1\\0\end{pmatrix} = \begin{pmatrix}\cos\phi\\ \sin\phi\end{pmatrix}, \quad R[\phi]f = R[\phi]\begin{pmatrix}0\\1\end{pmatrix} = \begin{pmatrix}-\sin\phi\\ \cos\phi\end{pmatrix}.$$

The matrix representing $R[\phi]$ has $R[\phi]e$, $R[\phi]f$ for its columns:

(4.18) $$R[\phi] = \begin{pmatrix}\cos\phi & -\sin\phi\\ \sin\phi & \cos\phi\end{pmatrix}.$$

MATRICES — TRANSFORMATIONS   115

4.8 Composite Transformations and Inverses

What happens if the plane is subjected first to a linear map M and subsequently to another linear map L? Their combined effect defines a new map called the *composite* of L and M, and denoted by LM: $LMv = L(Mv)$. The reason for using the symbolism of multiplication in composition will be justified below.

If L and M are linear, then

$$LM(u + v) = L(Mu + Mv) = LMu + LMv,$$

and

$$LMkv = LkMv = kLMv;$$

hence the composite LM is linear.

Let L, M, N be any three maps. We perform N followed by M followed by L: $v \to Nv \to M(Nv) \to L(M(Nv))$, and observe that $(LM)N$ and $L(MN)$ have precisely the same effect; $(LM)N$ may be interpreted as pausing between N and M, and $L(MN)$ as pausing between M and L, and such pauses do not affect the result of this triple composition. We conclude that composition is *associative*:

$$(LM)N = L(MN) = LMN.$$

Composition is associative for non-linear maps as well.

The transformation that maps every vector into itself is called the *identity* map and is denoted by I. It is clearly linear since

$$I(u + v) = u + v = Iu + Iv, \quad \text{and} \quad Iku = ku = kIu.$$

Its matrix representation has columns $Ie = e = \begin{pmatrix} 1 \\ 0 \end{pmatrix}$ and $If = f = \begin{pmatrix} 0 \\ 1 \end{pmatrix}$, so

(4.19) $$I = \begin{pmatrix} 1 & 0 \\ 0 & 1 \end{pmatrix}.$$

A map M is said to be *one-to-one onto* if the following two conditions are satisfied:

(i) For any vector v there is a vector u such that $Mu = v$. (onto)
(ii) If u and v are distinct vectors, so are Mu and Mv. (one-to-one)

A one-to-one onto map M has an *inverse* map M^{-1} which associates with each vector v its preimage, i.e. the vector sent to v by M. More explicitly, corresponding to every relation $Mu = v$, we write $M^{-1}v = u$. Then we note that

$$M^{-1}Mu = M^{-1}v = u$$

holds for every vector v. We conclude that $M^{-1}M$ and MM^{-1} are both equal to the identity map, so that

$$M^{-1}M = MM^{-1} = I.$$

If L is a linear one-to-one onto map, its inverse is linear. This is an easy consequence of the linearity of L and the definition of inverse. To prove it, we set $L(u) = r$, $L(v) = s$; then $L^{-1}r = u$ and $L^{-1}s = v$. Since L is linear,

$$L(u + v) = Lu + Lv = r + s.$$

Applying L^{-1} to this identity yields

$$L^{-1}(r + s) = u + v = L^{-1}r + L^{-1}s.$$

This establishes the additivity property (4.14) of L^{-1}. To demonstrate (4.15), note that

$$L^{-1}kr = L^{-1}kLu = L^{-1}Lku = ku = kL^{-1}r.$$

More generally, let us see under what conditions a linear mapping

$$M = \begin{pmatrix} a & c \\ b & d \end{pmatrix}$$

has an inverse; that is, under what conditions it is one-to-one onto. Let v be any vector. Is there a vector

$$u = \begin{pmatrix} x \\ y \end{pmatrix}$$

such that

$$Mu = Mxe + Myf = xMe + yMf = v?$$

Geometrically, this means: can we represent v as diagonal of a parallelogram with sides in the directions Me and Mf? See Figure 4.10. This is possible for all v if and only if the vectors Me and Mf are *not*

MATRICES—TRANSFORMATIONS

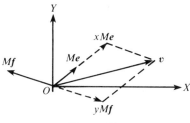

Figure 4.10

collinear, that is, if their components $\binom{a}{c}$ and $\binom{b}{d}$ are not proportional. Thus M has an inverse if and only if

$$ad - bc \neq 0.$$

Suppose $z = \binom{s}{t}$ is a vector such that $Mz = o = sMe + tMf$. If Me and Mf are not collinear, this implies that $s = t = 0$. Thus a matrix whose elements satisfy $ad - bc \neq 0$ maps only the zero vector to the zero vector. It follows that distinct vectors have distinct images (condition (ii) above); for if $Mu = Mv$, then $Mu - Mv = M(u - v) = o$, and so $u - v = o$ and $u = v$. We conclude that M has an inverse M^{-1} if and only if $ad - bc \neq 0$. We leave it to the reader to verify that

$$M^{-1} = \begin{pmatrix} d/D & -b/D \\ -c/D & a/D \end{pmatrix} \qquad \text{where } D = ad - bc.$$

4.9 Composition and Matrix Multiplication

Since we can represent linear mappings by matrices, let us find the matrix for the composite LM of two linear mappings M and L whose matrices are known:

(4.20) $$M = \begin{pmatrix} a & b \\ c & d \end{pmatrix}, \qquad L = \begin{pmatrix} g & h \\ k & l \end{pmatrix}.$$

We obtain the first column of the matrix LM as the image of e under the mapping M followed by L:

First column of $LM = L(Me) = L\binom{a}{c} = \begin{pmatrix} g & h \\ k & l \end{pmatrix}\binom{a}{c} = \binom{ga + hc}{ka + lc}$

by (4.17′) and the remark following (4.17′). Similarly,

Second column of $LM = L(Mf) = L\begin{pmatrix} b \\ d \end{pmatrix} = \begin{pmatrix} g & h \\ k & l \end{pmatrix}\begin{pmatrix} b \\ d \end{pmatrix} = \begin{pmatrix} gb + hd \\ kb + ld \end{pmatrix}$.

Thus

(4.21) $\qquad LM = \begin{pmatrix} g & h \\ k & l \end{pmatrix}\begin{pmatrix} a & b \\ c & d \end{pmatrix} = \begin{pmatrix} ga + hc & gb + hd \\ ka + lc & kb + ld \end{pmatrix}$.

In words: The entry a_{ij} in the ith row and jth column of the *product LM* of two matrices M, L is the dot product of the ith row of L with the jth column of M. Note that L and M play distinct roles in this product; composition of mappings and hence also multiplication of matrices is *not* commutative except in special cases. (Composition of non-linear mappings is also not commutative in general.)

Example: Let

$$L_1 = \begin{pmatrix} 1 & 2 \\ 3 & 4 \end{pmatrix}, \quad L_2 = \begin{pmatrix} 2 & 0 \\ 1 & -1 \end{pmatrix}.$$

Performing L_1 followed by L_2 corresponds to the matrix

$$L_2 L_1 = \begin{pmatrix} 2 & 0 \\ 1 & -1 \end{pmatrix}\begin{pmatrix} 1 & 2 \\ 3 & 4 \end{pmatrix} = \begin{pmatrix} 2\cdot 1 + 0\cdot 3 & 2\cdot 2 + 0\cdot 4 \\ 1\cdot 1 - 1\cdot 3 & 1\cdot 2 - 1\cdot 4 \end{pmatrix}$$

$$= \begin{pmatrix} 2 & 4 \\ -2 & -2 \end{pmatrix}.$$

Performing L_2 followed by L_1 corresponds to

$$L_1 L_2 = \begin{pmatrix} 1 & 2 \\ 3 & 4 \end{pmatrix}\begin{pmatrix} 2 & 0 \\ 1 & -1 \end{pmatrix} = \begin{pmatrix} 1\cdot 2 + 2\cdot 1 & 1\cdot 0 - 2\cdot 1 \\ 3\cdot 2 + 4\cdot 1 & 3\cdot 0 - 4\cdot 1 \end{pmatrix}$$

$$= \begin{pmatrix} 4 & -2 \\ 10 & -4 \end{pmatrix}.$$

The *sum* of two linear transformations L and M is defined by $(L + M)v = Lv + Mv$. There is a *zero* transformation (which maps all vectors into o); and the *negative* $-M$ of a transformation M is such that $Mv + (-M)v = o$. Each of these properties translates into a property of matrices. For example, for M, L defined in (4.20), $M + L$ sends e into $Me + Le = \begin{pmatrix} a + g \\ c + k \end{pmatrix}$ and f into $Mf + Lf = \begin{pmatrix} b + h \\ d + l \end{pmatrix}$ so that the entries of the matrix $M + L$ are the sums of corresponding entries of M and of L. The zero matrix has all its entries 0, and the entries of $-M$ are the negatives of the entries of M. To complete the algebra of matrices, we invite the reader to verify that addition is commutative and associative, and that matrix multiplication distributes over addition, i.e. that $L(M + N) = LM + LN$.

4.10 Rotations and the Addition Formulas of Trigonometry

We now return to rotation $R[\phi]$ by an angle ϕ, with the matrix given in (4.18) and ask: what is the effect of a rotation through ϕ followed by a rotation through θ? Composition of these transformations leads to the product $R[\theta]R[\phi]$ of the matrices representing rotations through ϕ and through θ. We compute their product by the rule (4.21):

$$R[\theta]R[\phi] = \begin{pmatrix} \cos\theta & -\sin\theta \\ \sin\theta & \cos\theta \end{pmatrix} \begin{pmatrix} \cos\phi & -\sin\phi \\ \sin\phi & \cos\phi \end{pmatrix}$$

$$= \begin{pmatrix} \cos\theta\cos\phi - \sin\theta\sin\phi & -\cos\theta\sin\phi - \sin\theta\cos\phi \\ \sin\theta\cos\phi + \cos\theta\sin\phi & -\sin\theta\sin\phi + \cos\theta\cos\phi \end{pmatrix}.$$

On the other hand, rotating through ϕ, then through θ, means rotating through $\phi + \theta$ and is represented by the matrix

$$R[\theta + \phi] = \begin{pmatrix} \cos(\phi + \theta) & -\sin(\phi + \theta) \\ \sin(\phi + \theta) & \cos(\phi + \theta) \end{pmatrix}.$$

Since identical linear maps must have identical matrices, we conclude that

(4.22) $$R[\theta]R[\phi] = R[\theta + \phi].$$

Equating corresponding entries, we obtain

(4.23)
$$\cos(\theta + \phi) = \cos\theta\cos\phi - \sin\theta\sin\phi,$$
$$\sin(\theta + \phi) = \sin\theta\cos\phi + \cos\theta\sin\phi,$$

the familiar addition formulas proved in trigonometry. Formula (4.22) also tells us that composition of rotations of the plane is commutative: $R[\theta]R[\phi] = R[\phi]R[\theta]$.

Clearly, the inverse of a rotation through an angle ϕ is a rotation through $-\phi$: $R^{-1}[\phi] = R[-\phi]$. To check our intuition, let us compute

120 MATHEMATICS IN SCIENCE

$R[\phi]R[-\phi]$ and ascertain that we get the identity:

$$R[\phi]R[-\phi] = \begin{pmatrix} \cos\phi & -\sin\phi \\ \sin\phi & \cos\phi \end{pmatrix} \begin{pmatrix} \cos(-\phi) & -\sin(-\phi) \\ \sin(-\phi) & \cos(-\phi) \end{pmatrix}$$

$$= \begin{pmatrix} \cos\phi\cos(-\phi) - \sin\phi\sin(-\phi) & -\cos\phi\sin(-\phi) - \sin\phi\cos(-\phi) \\ \sin\phi\cos(-\phi) + \cos\phi\sin(-\phi) & -\sin\phi\sin(-\phi) + \cos\phi\cos(-\phi) \end{pmatrix}.$$

Since $\cos(-\phi) = \cos\phi$ and $\sin(-\phi) = -\sin\phi$, we find that

$$R[\phi]R[-\phi] = \begin{pmatrix} \cos^2\phi + \sin^2\phi & \cos\phi\sin\phi - \sin\phi\cos\phi \\ \sin\phi\cos\phi - \cos\phi\sin\phi & \sin^2\phi + \cos^2\phi \end{pmatrix}$$

(4.24)

$$= \begin{pmatrix} 1 & 0 \\ 0 & 1 \end{pmatrix} = I.$$

Indeed, a rotation through ϕ and then through $-\phi$ is equivalent to a rotation through zero degrees, i.e., no rotation at all.

4.11 Reflections

We now investigate another class of mappings: reflections in a line l containing the origin. The reflection (or mirror image) P' of a point P in the line l is obtained by drawing the perpendicular PL from P to l and extending it to the point P' such that $LP' = PL$; see Figure 4.11(a). This figure illustrates a symmetry with respect to the line l; and since we have used R for rotation, we cannot also use it for reflection, so we use S, thereby reminding ourselves of symmetry.

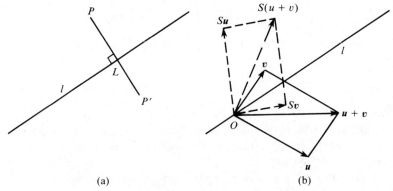

Figure 4.11

MATRICES — TRANSFORMATIONS

We claim that the reflection S in a line through 0 is a linear mapping. For clearly the reflection of a parallelogram is a parallelogram, and in particular, the diagonal $u + v$ of a parallelogram with sides u, v is mapped into the reflected parallelogram; i.e.

$$S(u + v) = Su + Sv,$$

see Figure 4.11(b). Similarly, the mirror image of a multiple of v is that multiple of the mirror image of v,

$$S(kv) = kS(v).$$

Having established the linearity of S, let us find its matrix representation. Suppose line l, through 0, makes an angle α with the unit vector e; see Figure 4.12. Denote by $S[\alpha]$ reflection in l. The columns of the matrix for $S[\alpha]$ are $S[\alpha]e$ and $S[\alpha]f$. Since e must be turned through α to coincide with l, it must be turned through 2α to yield its mirror image:

$$S[\alpha]e = \begin{pmatrix} \cos 2\alpha \\ \sin 2\alpha \end{pmatrix}.$$

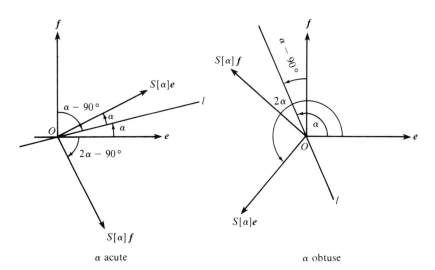

α acute $\qquad\qquad\qquad$ α obtuse

Figure 4.12

Since f is at 90° with e, f makes the angle $\alpha - 90°$ with l, and so the mirror image of f makes the angle $\alpha + (\alpha - 90°) = 2\alpha - 90°$ with e.

Hence
$$S[\alpha]f = \begin{pmatrix} \cos(2\alpha - 90°) \\ \sin(2\alpha - 90°) \end{pmatrix} = \begin{pmatrix} \sin 2\alpha \\ -\cos 2\alpha \end{pmatrix}.$$

We conclude that

(4.25) $$S[\alpha] = \begin{pmatrix} \cos 2\alpha & \sin 2\alpha \\ \sin 2\alpha & -\cos 2\alpha \end{pmatrix}$$

is the matrix representing reflection in the line l: $y = (\tan \alpha)x$.

What is the reflection, in the same mirror, of the reflection of a figure F? Clearly, it is F itself. Indeed, the composition of $S[\alpha]$ with itself yields

$$S[\alpha]S[\alpha] = S^2[\alpha] = \begin{pmatrix} \cos 2\alpha & \sin 2\alpha \\ \sin 2\alpha & -\cos 2\alpha \end{pmatrix}\begin{pmatrix} \cos 2\alpha & \sin 2\alpha \\ \sin 2\alpha & -\cos 2\alpha \end{pmatrix}$$

$$= \begin{pmatrix} \cos^2 2\alpha + \sin^2 2\alpha & \cos 2\alpha \sin 2\alpha - \sin 2\alpha \cos 2\alpha \\ \sin 2\alpha \cos 2\alpha - \cos 2\alpha \sin 2\alpha & \sin^2 2\alpha + \cos^2 2\alpha \end{pmatrix}$$

$$= \begin{pmatrix} 1 & 0 \\ 0 & 1 \end{pmatrix} = I,$$

so $S[\alpha]$ is its own inverse.

Although geometrically, rotations are very different from reflections, there is a striking similarity between the matrix describing the rotation $R[2\alpha]$, see (4.18), and the matrix for $S[\alpha]$, see (4.25): one can be obtained from the other by changing the signs in the second column. This can be expressed neatly by the matrix relation

(4.26) $$S[\alpha] = R[2\alpha]\begin{pmatrix} 1 & 0 \\ 0 & -1 \end{pmatrix}.$$

There is an equally striking similarity between $S[\alpha]$ and $R[-2\alpha]$: one can be obtained from the other by changing the signs in its second row; and in matrix language, this is expressed by

(4.27) $$S[\alpha] = \begin{pmatrix} 1 & 0 \\ 0 & -1 \end{pmatrix}R[-2\alpha].$$

What is the geometric meaning of the matrix $\begin{pmatrix} 1 & 0 \\ 0 & -1 \end{pmatrix}$ appearing in (4.26) and (4.27)? It is $S[0]$, reflection in the X-axis, as can be seen by setting $\alpha = 0$ in (4.25). So we may rewrite relation (4.26) for any angle α as

(4.26′) $$S[\alpha] = R[2\alpha]S[0],$$

and (4.27) as

(4.27') $$S[\alpha] = S[0]R[-2\alpha].$$

The first says that a reflection $S[\alpha]$ in the line $y = (\tan \alpha)x$ can be accomplished by reflection in the X-axis followed by a *counterclockwise* rotation through 2α; the second says that $S[\alpha]$ can be accomplished by a *clockwise* rotation through 2α followed by reflection in the X-axis.

From equations (4.26') and (4.27') we conclude that $R[2\alpha] S[0] = S[0]R[-2\alpha]$; since α is an arbitrary angle, we may state that for any angle γ,

(4.28) $$R[\gamma]S[0] = S[0]R[-\gamma].$$

In words: Reflection in the X-axis followed by rotation through γ is equivalent to a rotation through $-\gamma$ followed by reflection in the X-axis; both are equivalent to the reflection $S[\gamma/2]$.

We now ask two questions:

(i) What is the net effect of a reflection $S[\beta]$ followed by a rotation through γ?

(ii) What is the net effect of a reflection $S[\beta]$ followed by another reflection $S[\delta]$?

Both questions can be answered with the help of the algebra of matrices developed above. We know the matrix representations of all these maps, so we can find the matrix for each desired composite by multiplying the appropriate pair of matrices, in the proper order. We suggest that the reader perform these computations, use the trigonometric addition formulas derived earlier, and interpret the result geometrically. Here we shall take a different track to answer the two questions posed above. Our method avoids matrix multiplication and trigonometry; instead it uses the rules of algebra governing the composition of mappings.

By (4.26') we have

$$S[\beta] = R[2\beta]S[0].$$

Multiplying by $R[\alpha]$ we get

(4.29) $$R[\alpha]S[\beta] = R[\alpha]R[2\beta]S[0]$$
$$= R[\alpha + 2\beta]S[0].$$

By (4.26') with α replaced by $(\alpha/2) + \beta$, we have

$$S[(\alpha/2) + \beta] = R[\alpha + 2\beta]S[0].$$

Hence from (4.29) we get

(4.30) $$R[\alpha]S[\beta] = S[(\alpha/2) + \beta].$$

In words: Reflection in a line that makes angle β with the X-axis followed by rotation through α is equivalent to a single reflection in the line that makes angle $(\alpha/2) + \beta$ with the X-axis.

Next we tackle Question (ii), concerning the effect of two reflections: By (4.26'), we have

$$S[\delta] = R[2\delta]S[0].$$

By (4.27'),

$$S[\beta] = S[0]R[-2\beta].$$

Hence

$$S[\delta]S[\beta] = R[2\delta]S[0]S[0]R[-2\beta]$$
$$= R[2\delta]R[-2\beta]$$
$$= R[2(\delta - \beta)].$$

So

(4.31) $$S[\delta]S[\beta] = R[2(\delta - \beta)].$$

In words: Reflection in a line that makes angle β with the X-axis followed by reflection in the line that makes angle δ with the X-axis can be accomplished by the single rotation through the angle $2(\delta - \beta)$, which is twice the angle between our two mirrors.

We ask the reader to confirm that matrix multiplication gives the same answers to questions (i) and (ii). Geometric constructions offer a third way of deriving these results.

4.12 Rigid Motions (Isometries)

We have an intuitive feeling that rotations and reflections, as well as the displacements discussed earlier, preserve all geometric properties of figures such as size and shape. So, of course, does any composite of these three kinds of maps. All such maps are called variously *congruences*, *Euclidean motions*, or *rigid motions*.

We shall show that congruences can be characterized by a single geometric property. A mapping M is called *distance-preserving* if for any pair of points P and Q of the plane, the distance $d(P, Q)$ between P and

Q is equal to the distance between the images $M(P)$ and $M(Q)$:

(4.32) $$d(P,Q) = d(M(P), M(Q)).$$

A distance-preserving map is also called an *isometry*. Clearly, the composite of any two isometries is an isometry. Since displacement, rotation and reflection are obviously isometries, so are their composites.

We shall show now that every isometry of the plane is a congruence; more precisely that it can be obtained either by a rotation followed by a displacement or by a reflection followed by a displacement. Our arguments here will be entirely geometric and are based on the following fact, which we prove first:

A distance-preserving map which fixes two points is either the identity or a reflection in the line through the fixed points.

Let M be a distance-preserving map, and P, Q two fixed points, i.e. points which coincide with their images under M:

$$M(P) = P, \qquad M(Q) = Q.$$

Let R be any point not on the line through P and Q. Where is its image under M? Since M preserves distances, the sides of triangle PQR have the same lengths as those of the triangle with vertices $M(P)$, $M(Q)$, $M(R)$ (see Figure 4.13). Hence these triangles are congruent (*sss*). They share the side PQ, so $M(R)$ can have two possible positions:
(i) $M(R) = R$
(ii) $M(R)$ is the mirror image R' of R in the line PQ.

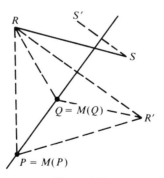

Figure 4.13

In case (i), M is the identity. To see this, take any fourth point S not on the line PQ. By the argument used above for R, we see that $M(S)$

either coincides with S or is the reflection S' of S in PQ; but it cannot be S' because $d(S', R) \neq d(S, R)$ for all R not on PQ. Thus distance is preserved only if $M(S) = S$. So in case (i), $M = I$.

Similarly, in case (ii), if M reflects R across the mirror PQ, its distance-preserving property forces it to reflect all points across this mirror. This concludes the proof of the italicized assertion.

Now suppose that N is any distance-preserving map of the plane. N maps the origin into some point $N(O)$. Denote by D the displacement along the segment from $N(O)$ to O, and form the composite M of the map N followed by D:

$$M = DN.$$

Since $M(O) = D(N(O)) = O$, the origin is a fixed point of M. Moreover, it is the composite of two isometries. Thus $M = DN$ is a distance-preserving map which fixes the origin of the plane. It suffices to study maps fixing O, since we can always reconstruct the original map N by following M by the displacement D^{-1}, the inverse of D: $D^{-1}M = D^{-1}DN = N$.

We consider next an isometry T which preserves the origin, i.e. $T(O) = O$, and show that T is either a rotation or a reflection in a line through the origin.

Let Q be any point other than the origin. Since T is an isometry, Q has the same distance from the origin as $T(Q)$. Therefore there is a rotation R about O which carries Q into $T(Q)$. Denote its inverse by R^{-1}. The composite $R^{-1}T$ of two isometries fixing O is an isometry fixing O. Moreover, $R^{-1}T$ also fixes Q (R^{-1} was chosen because it maps $T(Q)$ back into Q). Thus $R^{-1}T$ has two fixed points, O and Q. According to our preceding result, it is either the identity or a reflection in the line OQ.

If $R^{-1}T = I$, then $RR^{-1}T = T = R$, so T is a rotation.

If $R^{-1}T$ is reflection in the line OQ, we may write $R^{-1}T = S[\beta]$ where β is the angle OQ makes with the horizontal. Then $RR^{-1}T = T = RS[\beta]$. But we saw (see equation (4.30)) that a reflection followed by a rotation is equivalent to a single reflection.

We have now shown that every isometry of the plane is either a rotation followed by a displacement or a reflection followed by a displacement.[†] To summarize: An isometry that fixes 0 has one of two possible matrix representations:

$$(4.33) \quad R[\theta] = \begin{pmatrix} \cos\theta & -\sin\theta \\ \sin\theta & \cos\theta \end{pmatrix} \quad \text{or} \quad S[\phi/2] = \begin{pmatrix} \cos\phi & \sin\phi \\ \sin\phi & -\cos\phi \end{pmatrix}.$$

[†]A displacement can be thought of as a rotation through zero degrees followed by this displacement, and is therefore included in the first category.

4.13 Orthogonal Matrices

We observe next that an isometry that fixes O preserves the dot product of two vectors u, v. We consider the vector $u + v$ and its image $M(u + v)$. Since M is linear, $M(u + v) = M(u) + M(v)$, and since it preserves distances, $|M(u + v)|^2 = |u + v|^2$. The left side is equal to

$$(Mu + Mv) \cdot (Mu + Mv) = |Mu|^2 + 2Mu \cdot Mv + |Mv|^2$$
$$= |u|^2 + 2Mu \cdot Mv + |v|^2,$$

and the right side is equal to

$$(u + v) \cdot (u + v) = |u|^2 + 2u \cdot v + |v|^2.$$

It follows that

(4.34) $$Mu \cdot Mv = u \cdot v.$$

Recall that $u \cdot v = |u| \, |v| \cos \theta$, where θ is the angle between u and v. Denote the angle between Mu and Mv by ψ. Then, by (4.34), we have

$$\cos \psi = \cos \theta,$$

from which it follows that either $\psi = \theta$ or $\psi = -\theta$; i.e. isometries preserve the magnitudes of angles, though not necessarily their orientation.

In particular, an isometry maps orthogonal unit vectors onto orthogonal unit vectors. Therefore the columns of M (images of e and f) are orthogonal unit vectors. If Me must be rotated through $90°$ to coincide with Mf, M is a rotation; if it must be rotated through $-90°$ to coincide with Mf, M is a reflection.

Conversely, if the columns of a matrix

$$M = \begin{pmatrix} a & b \\ c & d \end{pmatrix}$$

are orthogonal unit vectors, i.e. if

$$a^2 + c^2 = b^2 + d^2 = 1 \quad \text{and} \quad ab + cd = 0,$$

then M is an isometry; for every unit vector may be written in the form

$$\begin{pmatrix} a \\ c \end{pmatrix} = \begin{pmatrix} \cos \theta \\ \sin \theta \end{pmatrix},$$

and there are two unit vectors orthogonal to it:

$$\begin{pmatrix} -\sin\theta \\ \cos\theta \end{pmatrix} \quad \text{and} \quad \begin{pmatrix} \sin\theta \\ -\cos\theta \end{pmatrix}.$$

If $\begin{pmatrix} b \\ d \end{pmatrix}$ is of the first kind, M is a rotation, and if $\begin{pmatrix} b \\ d \end{pmatrix}$ is of the second kind, M is a reflection. Matrices whose columns are orthogonal unit vectors are called *orthogonal matrices* and, as we have seen, represent isometries.

We observe in passing that the inverse of the rotation $R[\alpha]$ is the rotation $R[-\alpha]$, while the reflection $S[\beta]$ is its own inverse. In both cases, the matrices representing the inverses of these isometries can be obtained from the matrices representing them by interchanging the rows and columns, as can be seen from equations (4.18) and (4.25). The matrix obtained from a given matrix M by interchanging its rows with its columns is called the *transpose* M^* of M. Thus we see that isometries which fix the origin are represented by matrices whose inverses are their transposes. Since the inverse of an isometry is an isometry, it follows that the rows of an orthogonal matrix are orthogonal unit vectors, just as the columns are.

We shall leave it to the reader to show that the transpose M^* of any matrix M satisfies the following relation:

(4.35) $\qquad\qquad M\boldsymbol{u} \cdot \boldsymbol{v} = \boldsymbol{u} \cdot M^*\boldsymbol{v} \qquad$ for all vectors $\boldsymbol{u}, \boldsymbol{v}$.

Now if M is an isometry, it preserves the dot product:

$$M\boldsymbol{u} \cdot M\boldsymbol{v} = \boldsymbol{u} \cdot \boldsymbol{v} \qquad \text{for all vectors } \boldsymbol{u}, \boldsymbol{v}.$$

Applying (4.35) to the left side, we have $M\boldsymbol{u} \cdot M\boldsymbol{v} = \boldsymbol{u} \cdot \boldsymbol{v} = \boldsymbol{u} \cdot M^*M\boldsymbol{v}$; hence

$$\boldsymbol{u} \cdot \boldsymbol{v} - \boldsymbol{u} \cdot M^*M\boldsymbol{v} = 0 \quad \text{or} \quad \boldsymbol{u} \cdot (\boldsymbol{v} - M^*M\boldsymbol{v}) = 0 \quad \text{for all vectors } \boldsymbol{u}.$$

This says that the vector $\boldsymbol{v} - M^*M\boldsymbol{v}$ is orthogonal to all vectors in the plane, and this is true only for the zero vector; so

$$\boldsymbol{v} = M^*M\boldsymbol{v} \qquad \text{for all vectors } \boldsymbol{v}.$$

From this we conclude that $M^*M = I$, obtaining a second proof of the fact that the transpose of an orthogonal matrix is its inverse.

4.14 Coordinate Transformations

We now turn to the following problem: In analytic geometry, the coordinates of a point depend on the choice of axes. These are usually

MATRICES — TRANSFORMATIONS

chosen to be as convenient as possible for describing the situation of interest. Frequently, in order to describe a new situation, it is efficient to introduce new coordinate axes. The problem facing us then is to relate the coordinates of a point in the original system to its coordinates in the new system. We shall show how the matrix concept and the theory developed for it is a powerful tool for solving this problem.

Denote the original axes by X, Y, the new axes by $\overline{X}, \overline{Y}$, and suppose that the two systems have a common origin. We consider two cases:

(i) The $\overline{X}, \overline{Y}$ axes are obtained by rotating the old X, Y axes through some angle α, $0 \le \alpha < 360°$. See Figure 4.14.

(ii) The $\overline{X}, \overline{Y}$ system has an orientation opposite to that of the X, Y system; that is, a rotation through α brings the positive X-axis into coincidence with the positive $\overline{X}$-axis, but it brings the positive Y-axis into coincidence with the negative $\overline{Y}$-axis, see Figure 4.15.

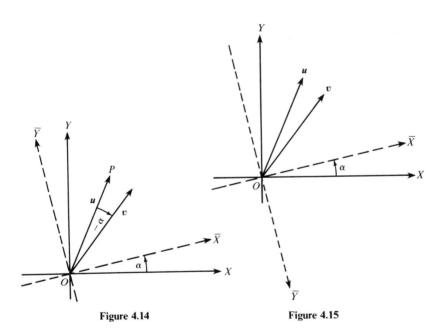

Figure 4.14   Figure 4.15

Now let P be any point in the plane, and denote by u the vector $\overline{OP}$. The crucial observation is that in case (i) the relation of u to the new $\overline{X}, \overline{Y}$ axes is the same as the relation to the old axes of the vector v obtained by rotating u through the angle $-\alpha$. In other words, the new components $\bar{x}$, $\bar{y}$ of u are the old components of $v = R[-\alpha]u$.

In case (ii), the $\bar{x}$-component of u is the same as in case (i), but the $\bar{y}$-component is the negative of the y-component we obtained in case (i).

To summarize: Under a rotation of the coordinate system, a vector $u = \begin{pmatrix} x \\ y \end{pmatrix}$ is transformed into

$$(4.36) \quad \bar{u} = \begin{pmatrix} \bar{x} \\ \bar{y} \end{pmatrix} = R \begin{pmatrix} x \\ y \end{pmatrix}, \quad \text{where} \quad R = \begin{pmatrix} \cos\alpha & \sin\alpha \\ -\sin\alpha & \cos\alpha \end{pmatrix};$$

and under a rotation and reflection of the coordinate system, u is transformed into

$$(4.37) \quad \bar{u} = \begin{pmatrix} \bar{x} \\ \bar{y} \end{pmatrix} = S \begin{pmatrix} x \\ y \end{pmatrix}, \quad \text{where} \quad S = \begin{pmatrix} \cos\alpha & \sin\alpha \\ \sin\alpha & -\cos\alpha \end{pmatrix}.$$

4.15 A Matter of Notation

Now that we know that every distance and origin preserving transformation of an X, Y system to an $\bar{X}, \bar{Y}$ system leads either to the component transformation (4.36) (direct isometry), or to the component transformation (4.37) (opposite isometry), we shall introduce some abbreviations for the elements of the orthogonal matrices R and S. We set $\cos\alpha = z$; then

$$\sin\alpha = \begin{cases} \sqrt{1-z^2} & \text{for } 0 \leq \alpha \leq 180° \\ -\sqrt{1-z^2} & \text{for } 180° < \alpha < 360°, \end{cases}$$

and the matrix R in (4.36) becomes

$$(4.38) \quad R = \begin{pmatrix} z & \sqrt{1-z^2} \\ -\sqrt{1-z^2} & z \end{pmatrix} \quad \text{or} \quad R' = \begin{pmatrix} z & -\sqrt{1-z^2} \\ \sqrt{1-z^2} & z \end{pmatrix},$$

while the matrix S in (4.37) becomes

$$(4.39) \quad S = \begin{pmatrix} z & \sqrt{1-z^2} \\ \sqrt{1-z^2} & -z \end{pmatrix} = \begin{pmatrix} 1 & 0 \\ 0 & -1 \end{pmatrix} R,$$

or

$$S' = \begin{pmatrix} z & -\sqrt{1-z^2} \\ -\sqrt{1-z^2} & -z \end{pmatrix} = \begin{pmatrix} 1 & 0 \\ 0 & -1 \end{pmatrix} R'.$$

MATRICES—TRANSFORMATIONS

Conversely, given any orthogonal matrix $\begin{pmatrix} A & B \\ C & D \end{pmatrix}$, i.e.

$$A^2 + B^2 = C^2 + D^2 = 1, \quad \text{and} \quad AB + CD = AC + BD = 0,$$

we can find an angle α, $0 \le \alpha < 360°$, such that

(i) for $AD - BC = 1$, we have $A = \cos \alpha$, $B = \sin \alpha (= \pm \sqrt{1 - A^2})$,

$$C = -\sin \alpha (= \mp \sqrt{1 - A^2}), \quad D = \cos \alpha; \quad \text{and}$$

(ii) for $AD - BC = -1$, $A = \cos \alpha$, $B = \sin \alpha (= \pm \sqrt{1 - A^2})$,

$$C = \sin \alpha (= \pm \sqrt{1 - A^2}), \quad D = -\cos \alpha.$$

Although we could have deduced equations (4.36), (4.37) even more succinctly, we prefer the present argument because it is echoed—albeit faintly—in a subsequent argument.

By considering orthogonal transformations in some detail and by emphasizing the rough analogy between nailing with and without hammers and doing algebra with and without matrices, we have tried to make it easier to appreciate the facility afforded by matrix techniques. But do not mistake analogy for mathematical appreciation; in particular, there is no substitute for working out and pondering over the mathematics of Chapter 4 for yourself.

CHAPTER FIVE

What is Time? Einstein's Transformation Problem

The basic problem of relativity theory arises out of trying to state with mathematical precision what we mean when we use the phrase "at the same time."

You will probably find the contents of this chapter more difficult to grasp than those of the preceding chapters not because they are mathematically more demanding, but because this chapter, and the next, require that you *unthink* certain firmly held notions.

Once we understand how the problem arose and why our old notions are inadequate for attacking it, we can appreciate how certain mathematical principles (including the matrix algebra of Chapter 4), used with bold imagination, lead to successful strategies.

5.1 The Michelson-Morley Experiment

A. A. Michelson (1852–1931), awarded the 1907 Nobel Prize for Physics, was one of the world's greatest experimental physicists. He is perhaps best introduced by the following anecdote: Asked by a father if his son should be encouraged to continue his studies to become a physicist, Michelson is said to have replied, "No, I advise your son not to study physics. It is a dead subject. What there is to know, we know—except that possibly we could measure a few things to the sixth decimal place instead of the fourth."

The irony of the story is that Michelson is the man whose experiments led to such a revolution that we have learned more about physics in the last sixty years than in all the preceding centuries.

WHAT IS TIME ?

But this story is revealing as well as ironical. Michelson was a man with a passion for accuracy, a man who measured everything to the sixth decimal place. In particular, he had, in the late 1870's, by most ingenious experimentation, measured the velocity of light with hitherto unheard-of accuracy. The velocity of the earth in its journey round the sun having been determined with fair accuracy from astronomical data, Michelson's next ambition was to remeasure it himself—to the sixth decimal place. With this objective in sight, in 1881, he made the experiment that was to make him famous.

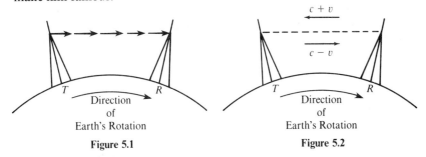

Figure 5.1   Figure 5.2

The concept upon which this experiment was based was simple. Suppose that we put a transmitter T and a receiver R a certain distance apart on the surface of the earth and measure the time taken for a signal, a flash of light, to go from T to R. See Figure 5.1. The signal sent from T with the enormous velocity of light[†] c has to overtake R, which is moving ahead with velocity v, the velocity of the earth. Therefore it has velocity $c - v$ relative to R and in consequence will take a little more time to reach R than it would if the earth were at rest. And if a return signal is sent from R back to T, it is approached by T as it approaches T, so that its velocity relative to T is $c + v$. See Figure 5.2. Therefore the return signal will take less time than it would if the earth were at rest and still less time than the initial, outgoing signal does.

Let us describe briefly the ingenious experimental arrangement to carry out this observation. See Figure 5.3. We have a light source L which sends a light ray in the direction of the motion of the earth. This ray falls on a mirror M_1 which is partially transparent and stands under an angle of 45° against the incoming ray. Thus, a part of the light is reflected under 90° to a mirror M_2 and a part goes through to a mirror M_3. Observe that the light ray has now been split up into two rays, one moving along the line $M_1 M_2$, that is, perpendicular to the earth's motion,

---

[†] The symbol c for velocity of light is customarily used; it is the first letter of the Latin word *celeritas*, meaning speed.

and the other moving along M_1M_3 parallel to the earth's motion. Both rays are then reflected at the mirrors M_2 and M_3 and return to the transparent mirror M_1. Now, part of the vertical light ray M_2M_1 passes through M_1 and goes to the objective of an interferometer J. At the same time a part of the light ray M_3M_1 is reflected at M_1 and also enters into the same interferometer. Thus, we mix in J the light of two different travel histories. The two types of light differ in their part by the difference in time which is necessary to travel from M_1 to M_2 and back as compared to the time which it takes to travel from M_1 to M_3 and back. You do not need to know the operation of an interferometer. It is sufficient to know that such an instrument is sensitive enough to compare light rays coming from the same origin but having spent different times in travel. Being mathematicians, we shall rather calculate the expected difference in travel time which the instrument will measure.

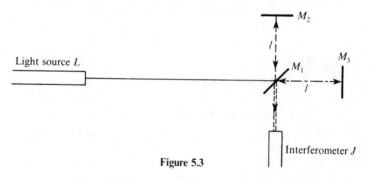

Figure 5.3

Let l be the distance between the mirrors M_1 and M_2, and M_1 and M_3 which, as you see, we assume to be equal. The travel time from M_1 to M_3 and back is evidently

$$\tau = \frac{l}{c-v} + \frac{l}{c+v} = \frac{2lc}{c^2 - v^2},$$

since in the forward motion light should travel with the lesser relative velocity $c - v$ and in return with the larger velocity $c + v$, as we discussed before.

It is more difficult to find the travel time from M_1 to M_2 and back. Let us look at the experiment from a point in outer space, so that we do not participate in the motion of the earth. At the moment when the ray leaves the mirror, M_1, this mirror has the position $M_1^{(1)}$ in space, and M_2 sits at the point $M_2^{(1)}$. But, suppose it takes the time $\frac{1}{2}t$ until the ray hits the mirror M_2. During this time the mirror M_2 has shifted in the direction of the earth's motion and sits at the point $M_2^{(2)}$ in space. It reflects the light

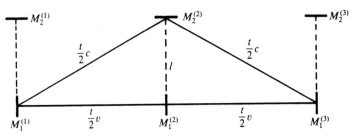

Figure 5.4

back to M_1; by reason of symmetry it will take the same time $\tfrac{1}{2}t$ to return from M_2 to M_1. But when the light reaches M_1, its position in space will be at $M_1^{(3)}$. We know that the distance $M_1^{(1)}$ to $M_1^{(3)}$ is given by vt since t is the total travel time and since the earth moves with the speed v. On the other hand, the light which travels with speed c had to cover in the time $\tfrac{1}{2}t$ the distance $M_1^{(1)}$ to $M_2^{(2)}$. Thus, in the right triangle $M_1^{(1)}M_1^{(2)}M_2^{(2)}$ all three sides are known as indicated in Figure 5.4. By the Pythagorean theorem, we have $l^2 = \tfrac{1}{4}t^2(c^2 - v^2)$, that is,

$$t = \frac{2l}{\sqrt{c^2 - v^2}}.$$

The travel time τ from M_1 to M_3 is not the same as that needed to go from M_1 to M_2. The ratio of travel times is

$$\frac{\tau}{t} = \frac{c}{\sqrt{c^2 - v^2}} = \frac{1}{\sqrt{1 - (v^2/c^2)}}.$$

The square root which occurs here in our elementary considerations is characteristic for relativity theory and will occur later in quite a different context.

The above reasoning allowed Michelson to predict a time difference in travel time and to adjust his instruments in such a way that the effect could be safely measured.

Although the idea behind this experiment is so simple, the refinements necessary to achieve the accuracy that Michelson demanded made the actual experimental set-up a hive of ingenuity. Michelson achieved such accuracy that he would have been able to determine v, the earth's velocity, even if it had been moving only one tenth as fast as it does.

Michelson made the measurement and created a scandal in physics. What value for v did he get? Zero. Yes, ZERO. The flash of light takes precisely the same time to go from T to R as from R to T. But this is preposterous. Even the small boy who steals apples from an orchard appreciates the importance of *relative velocity*—even before he can spell the words. He knows perfectly well that to escape a good hiding he must continue to run away from, not towards, the wrathful farmer hard in his pursuit. But surely there can be no difference in principle between being chased by a farmer and a flash of light? The flash is more fleet of foot, that's all.

Physicists could not believe their eyes. Michelson was forced to repeat the experiment in 1887, this time assisted by E. W. Morley. Again the answer was zero, against all understanding of physics. How, for goodness' sake, could the velocity of light relative to a moving object be the same when overtaking the object as when moving towards it? Despite heated discussion, the cold fact is that the velocity of light is invariant.

Yes, of course the velocity of light is invariant. OF COURSE? Well, suppose that the Michelson-Morley experiment made late in the nineteenth century had been made in the sixteenth century when Copernicus was arguing against the generally accepted Ptolemaic earth-centered theory. Everyone, everywhere, everywhen resists having his deepest beliefs uprooted. Copernicus's contemporaries, brought up in the self-gratifying supposition that their earth was the fixed central pivot of the planetary system, could not bring themselves to suppose that the sun had better claim to the coveted pivotal position. Human nature being human nature, understandably Copernicus encountered not only strong but also bitter opposition. Understandably his opponents would have immediately heralded the Michelson-Morley experiment as a conclusive demonstration that our earth does not move.

But with time, volte-face. Even by the time of Newton, scientists were trying to devise experiments that would demonstrate the contrary—that our earth does indeed move. But by this time there was the contrary belief to protect. With the tremendous success of Newton's mechanics when applied to Copernicus's sun-centered theory, the sun, not the earth, had to be the central pivot. And so, when the Michelson-Morley experiment was accepted in 1888, by then with the sun-centered system even more deeply rooted in scientific thought, that our earth does not move was the one explanation of the experimental result to which no one gave as much as a passing thought.

5.2 What Time Is It?

After a discussion of the Michelson-Morley experiment and its conceivable consequences had been prolonged in scientific journals for some twenty years, Einstein came up with a penetrating remark. "What," he asked, "do we mean by saying that two events happened *at the same time*.† How do we know that everybody can agree what the time is at this very instant?"

In this age of jet travel it is a commonplace experience that different longitudes have different times. A telephone call from San Francisco to New York immediately confirms a different clock reading there. This communication, made by electricity at the speed of light, is so rapid that for all practical purposes the West Coast inquirer hears the East Coast answerer's reply at the same time as it is spoken in New York. At the same time the two clocks record different times, yet neither clock is wrong. Somehow or other clock readings are dependent upon an agreement about how we measure time. This remark is silly or subtle, as you please. Doesn't it sound peculiar to say that at the *same time* different clocks can *correctly* record *different times*? We call to mind the visiting philosophy professor who, in concluding his discourse on *Time* with the remark, "So you see, gentlemen, I do not know what Time is" looked at his watch—and dashed to catch his train.

Even if we dismiss the-different-times-at-the-same-time paradox as merely verbal, it is none the less a fact that, with interplanetary travel a realistic probability, the business of synchronising clocks becomes of practical importance. And cosmic voyaging introduces a complication not encountered in terrestrial travel. Whereas the time lag in hearing in San Francisco what is said in New York is about one fiftieth of a second, the time lag in interplanetary communication (by radio waves with the velocity of light) is a matter of minutes, and that between the earth and the stars, months.

Suppose, for example, that a radio signal sent by E, an observer on earth, to A, an astronaut in outer space $60 \times 186,000$ miles away, takes 1 minute. If E sends his signal when his clock records 12 o'clock, then his signal reaches A when E's watch records 12:01. A, in receiving the signal, sets his clock at 12:01. To confirm receipt of E's signal A immediately signals back. And since the distance between A and E remains unchanged, this return signal also takes 1 minute. Therefore E receives A's acknowledgment at 12:02 by E's clock.

This, you will say, is all very simple. Surely there's no difficulty. At 12:01 E says to himself, "A is now receiving my signal sent at 12:00 by my clock and setting his clock to read 12:01, the same time as mine." And from E's point of view isn't his conviction established beyond doubt by his clock reading 12:02 when he receives A's return signal?

---

†This question was raised in 1898 by Poincaré, see *Subtle is the Lord* by A. Pais, p. 133; Einstein was familiar with Poincaré's speculations.

The point is that whereas E *knows* that he himself sends a signal at 12:00 by his clock and *knows* he receives A's confirmatory signal at 12:02, E does *not* know that A received his (E's) signal when his (E's) clock read 12:01. Oh yes, he is convinced, but he does not know. He cannot be in both places at once to find out. *He has no method of direct verification.*

To synchronize clocks by means of light or radio signals, we must know the velocity with which our signals are transmitted, but to determine this velocity, we must know how long the transmission takes. To attempt to synchronize clocks without knowing the velocity of light and to determine the velocity of light without using a clock is just as futile as to try to produce hens without eggs and eggs without hens.

It is arguable that if eyewitnesses to E's signalling A are separated by great distances, then they must report vastly different, yet equally reliable, opinions of the time indicated by E's clock when his signal reaches A. All very confusing. To go into great detail is to invite great confusion. Physicists went into very great detail. Many on-paper experiments were made in which people frantically set their watches as they hastily got on and off trains, trams, boats, and bicycles, scheduled for immediate departure at velocities near that of light. Scientific journals were full of these wild excursions.

Of course, it is easy to poke fun. The physicists were able, serious-minded people, trying to figure out an important problem. Their real difficulty was conceptual rather than mathematical; quite literally, they didn't know what they were talking about. Whereas we all know well enough how to use the concept of time in everyday conversation, we are at a loss when we come to map its logical geography.

5.3 Einstein's Space-Time Transformation Problem

The matter was finally cleared up in 1905 by Einstein. He did two things of the greatest possible importance for physics: (1) He saw more clearly than any of his contemporaries what the basic problem is and gave it precise mathematical formulation; (2) he solved it. The first is by far the more difficult achievement. We shall devote this section to (1), the next to (2).

Although Einstein (1879–1955) was an imaginative thinker with his head in the clouds, he had both feet on the ground. He did not spend several years in the Swiss Patents Office for nothing. A professor of metaphysics would have asked: "What is Time?" or, "What is the essence of *Time*?" or more recently, "What do we mean by *Time*?"

WHAT IS TIME ?

Einstein, on the contrary, asked: "If a happening is observed by two people, how are one person's answers to the questions, 'Where', 'When?', related is the other's?" He looked for an answer in terms of measuring rods and clocks, not essences or semantics. He was a professor of physics, not metaphysics.

Before describing precisely how two people's observations are related to each other, we specify in detail how each single person records an observation. First of all, each observer is equipped with a *standard measuring rod* and a *standard clock*; by "standard" we mean issued by the same supply depot and certified to be identical.

The events to be recorded are flashes of light occurring in the darkness of night. An observation consists of giving the *location* and the *time* of the event. For example, suppose an observer sees a flash of light originating next to a familiar landmark; its position relative to the observer, denoted by x, can easily be measured with the standard rod in the light of the day. When the observer sees the flash, he records the time t_f from his standard clock. He knows his distance $|x|$ from the location of the flash, and he knows that light travels with velocity c. He can therefore compute the time t at which the event occurred by subtracting from t_f the time $|x|/c$ it took the signal to cover the distance $|x|$ between him and the point where the flash originated:

$$t = t_f - |x|/c.$$

Even if the flash occurs at an unknown distance from the observer he can record the event with the help of an assistant; see Exercise 5.1.

Exercise 5.1. An observer and his assistant, their standard clocks synchronized, are at their observation posts O and A, respectively, separated by the known distance l; see Figure 5.5. A flash occurs at some point between them, call it F. Each records his clock reading t_O and t_A respectively, when he sees the flash. Find the location and time, (x, t), of the event observed by them.

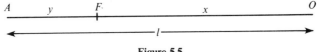

Figure 5.5

[Hint: Let x and y denote the distances FO and FA, respectively. From the relations

$$x + y = l, \qquad t_O = t + \frac{x}{c}, \qquad t_A = t + \frac{y}{c},$$

x and t can be found.]

Now consider two observers; each is at rest on his own observation platform, and each platform moves at constant velocity relative to the other. Say the first moves with velocity v relative to the second, so the second moves with velocity $-v$ relative to the first. Before these observers can compare their observations, they must make sure that they measure time from the same reference point. They can accomplish this by setting their clocks at 0 at the instant when they pass each other. Each describes the location of an event by its distance from himself.

As a homely example of two such observers, imagine a station master S at rest in the station house of a railroad, and imagine a motorman M at rest in his cab of a train moving at constant speed v with respect to the stationmaster. As the train thunders along, the clocks of S and M are synchronized at 0 the instant M passes S. For any event, each will record two numbers: a distance and a time. We denote by (x, t) those recorded by the station master, and by $(\bar{x}, \bar{t})$ those recorded by the motorman. For example, suppose a powerful stroke of lightning strikes a tree at F and both observers see it in a very brief moment of illumination. Both report the event that the tree has been struck. Observer S records that the event took place at a distance x from him and at a time t that he calculated by having looked at his watch when he saw the tree and by subtracting the time x/c which it took the light to travel from F to him (see Figure 5.6). The motorman M, noting that the tree, when struck, was alongside the middle of the dining car of his train, and keenly conscious of the Michelson-Morley experiment, makes a similar calculation using his own watch and states that the event took place at a distance $\bar{x}$ from him and at a time $\bar{t}$.

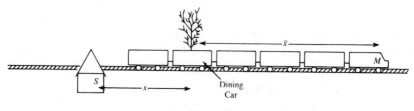

Figure 5.6

Einstein's question is: How are the answers to the questions "where?", "when?" of two such observers of the same event related? Or, in a mathematical formula, what are the functions relating x, t to $\bar{x}, \bar{t}$?

Certainly, these functions will depend on the velocity v of one platform relative to the other. Let us denote them by

(5.1) $$\bar{x} = F(x, t; v), \qquad \bar{t} = G(x, t; v).$$

WHAT IS TIME ?

To find out what kind of functions F and G are, we perform this thought experiment: We introduce an assistant station master A standing alongside the track to the left of the station master S and at a distance a from him. We also introduce a brakeman B who rides in the last car of the train, $\bar{a}$ units behind the motorman's seat, as measured on the train. See Figure 5.6.' Now an event observed by the station master at location x will be recorded by his assistant as having occurred at the location

(5.2) $$x_A = x + a.$$

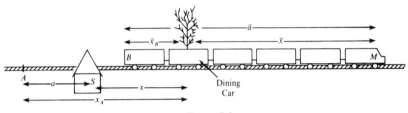

Figure 5.6'

The observations of the brakeman and motorman, $\bar{x}_B$ and $\bar{x}$ respectively, are similarly related:

(5.3) $$\bar{x}_B = \bar{x} + \bar{a}.$$

The brakeman and assistant station master set their clocks at 0 when they pass each other. The time recorded as t by the station master will be recorded as t_A by his assistant. Since they have identical clocks, these clocks readings differ by some constant s, so

(5.4) $$t_A = t + s.$$

Similarly, the times $\bar{t}$ and $\bar{t}_B$ recorded by motorman and brakeman, respectively, are related by

(5.5) $$\bar{t}_B = \bar{t} + \bar{s}.$$

Surely, the relation between the observations made by observers A and B are described by the *same functions* F and G that govern the relation between the observations made by S and M. Therefore

(5.6) $$\bar{x}_B = F(x_A, t_A; v)$$
$$\bar{t}_B = G(x_A, t_A; v).$$

Since we are considering the same two platforms, one moving with fixed velocity v relative to the other, we shall not explicitly indicate the dependence of F and G on v in the rest of this discussion.

When the motorman passes the station master, they both record $x = \bar{x} = 0$, $t = \bar{t} = 0$. How is the same event recorded by the assistant and the brakeman? Setting $x = 0$ in (5.2) and $t = 0$ in (5.4) shows that A records that this event occurred at $x_A = a$ and time $t_A = s$. Similarly, it follows from (5.3) and (5.5) that B records it as having occurred at $\bar{x}_B = \bar{a}$ and time $\bar{t}_B = \bar{s}$. It follows therefore from (5.6) that

(5.7) $$\bar{a} = F(a, s), \qquad \bar{s} = G(a, s).$$

Now, we combine (5.3) and (5.5) with (5.6):

$$\bar{x} + \bar{a} = \bar{x}_B = F(x_A, t_A)$$
$$\bar{t} + \bar{s} = \bar{t}_B = G(x_A, t_A).$$

We express x_A and t_A from (5.2) and (5.4):

(5.8)
$$\bar{x} + \bar{a} = F(x + a, t + s)$$
$$\bar{t} + \bar{s} = G(x + a, t + s).$$

Now use (5.1) to express $\bar{x}$, $\bar{t}$, and (5.7) to express $\bar{a}$, $\bar{s}$ on the left in (5.8); we get

(5.9)
$$F(x, t) + F(a, s) = F(x + a, t + s)$$
$$G(x, t) + G(a, s) = G(x + a, t + s).$$

Since the brakeman and the assistant can be positioned anywhere, (5.9) holds for any a and s. These relations show that F and G are *linear functions*. Such functions are of the form

(5.10)
$$\bar{x} = F(x, t) = Ax + Bt,$$
$$\bar{t} = G(x, t) = Cx + Dt.$$

We deduce now yet another property of the relation between x, t and $\bar{x}$, $\bar{t}$ that will enable us to determine the constants A, B, C, D as functions of v.

Proposition: If

(5.11) $$|x| = c|t|,$$

then

(5.11)′ $$|\bar{x}| = c|\bar{t}|.$$

WHAT IS TIME ?

Proof: We let t be negative; we recall from p. 139 that $t = t_f - |x|/c$. If (5.11) holds and t is negative, then $t = -|x|/c$; it follows that t_f is zero. We recall that at time zero both observers are in the same place; therefore the second observer also observes the event when *his* clock reads zero. He will then assign the time $\bar{t} = -|\bar{x}|/c$ to the event. This shows that (5.11)' holds.

So far we have assumed that t is negative; but we have already shown that (x, t) and $(\bar{x}, \bar{t})$ are linearly related. It follows that $(-x, -t)$ corresponds to $(-\bar{x}, -\bar{t})$. This proves the proposition.

We can restate the proposition in a form that avoids the absolute value sign:

(5.12) $\quad$ If $x^2 - c^2 t^2 = 0$, then $\bar{x}^2 - c^2 \bar{t}^2 = 0$.

Since $\bar{x}, \bar{t}$ are linear functions of x and t, $\bar{x}^2 - c^2 \bar{t}^2$ is a quadratic function of x and t. Now if one quadratic function in x and t is zero whenever another quadratic function in x and t is zero, then both have two linear factors in common (by the factor theorem of algebra) and hence one is a multiple of the other. Thus

(5.13) $\quad\quad \bar{x}^2 - c\bar{t}^2 = e(v)(x^2 - c^2 t^2).$

As we have indicated, the factor e may depend on the velocity v with which the second platform is moving with respect to the first.

When $v = 0$, the two platforms are at rest with respect to each other, therefore $\bar{x} \equiv x$, $\bar{t} \equiv t$. It follows that

(5.14) $\quad\quad\quad\quad\quad\quad e(0) = 1.$

The first platform is moving with velocity $-v$ when observed from the second platform. Therefore we obtain from (5.13), upon interchanging the roles of x, t and $\bar{x}, \bar{t}$, that

$$x^2 - c^2 t^2 = e(-v)(\bar{x}^2 - c^2 \bar{t}^2).$$

Comparing this with (5.13) we deduce that

(5.15) $\quad\quad\quad\quad\quad e(-v)e(v) = 1.$

A law of physics should not depend on whether the motion is to the right or left (this is yet another symmetry principle); therefore

$$e(-v) = e(v).$$

Setting this into (5.15) gives $e^2(v) = 1$; therefore $e(v) = 1$ or $e(v) = -1$.

But according to (5.14) $e = 1$ at $v = 0$, and since e is surely a continuous function of v, we conclude that $e(v) = 1$ for all v! Setting this into (5.13) we get the relation

$$(5.16) \qquad \bar{x}^2 - c^2\bar{t}^2 = x^2 - c^2t^2.$$

This is the *fundamental law of the special theory of relativity*.

5.4 Einstein's Solution

Einstein's problem of finding all linear transformations (5.10)

$$\bar{x} = Ax + Bt, \qquad \bar{t} = Cx + Dt$$

subject to $\bar{x}^2 - c^2\bar{t}^2 = x^2 - c^2t^2$ reminds us of the problem of finding all distance preserving maps solved in Chapter 4. The analogy becomes even more striking if we set

$$(5.17) \qquad ct = y, \qquad c\bar{t} = \bar{y}$$

in equations (5.10) and (5.16); for then Einstein's problem reads: Find all transformations

$$(5.18) \qquad \begin{aligned} \bar{x} &= Ax + Ey \\ \bar{y} &= Fx + Dy \end{aligned} \quad \text{i.e.} \quad \begin{pmatrix} \bar{x} \\ \bar{y} \end{pmatrix} = \begin{pmatrix} A & E \\ F & D \end{pmatrix} \begin{pmatrix} x \\ y \end{pmatrix}$$

[where E and F are the abbreviations

$$(5.19) \qquad E = B/c, \qquad F = cC, \qquad c = \text{speed of light}]$$

subject to

$$(5.20) \qquad \bar{x}^2 - \bar{y}^2 = x^2 - y^2.$$

We use (5.18) to compute

$$\bar{x}^2 - \bar{y}^2 = (Ax + Ey)^2 - (Fx + Dy)^2$$
$$= (A^2 - F^2)x^2 + 2(AE - FD)xy + (E^2 - D^2)y^2.$$

This is $x^2 - y^2$ if

$$(5.21) \qquad A^2 - F^2 = 1, \qquad D^2 - E^2 = 1, \qquad AE - FD = 0.$$

WHAT IS TIME ?

These are three equations for four quantities. The count is right, since we expect a one-parameter family of solutions, in analogy with the one-parameter family of rotations which solved the problem of finding all direct isometries in Chapter 4. In that case, the parameter was the angle of rotation; in the present case, the parameter will be the relative velocity v of one platform with respect to the other.

From the third equation in (5.21) we deduce that $A/F = D/E = k$, so $A = kF$, $D = kE$. Setting these into the first and second equations, we have $(k^2 - 1)F^2 = (k^2 - 1)E^2 = 1$; so $E^2 = F^2$, and therefore $A^2 = D^2$. It follows that

(5.22)   either $A = D$ and $F = E$ or $A = -D$ and $F = -E$.

Either choice satisfies the third equation in (5.21) and reduces the other two to a single equation $A^2 - F^2 = 1$ whose solution is

(5.23) $$A = \sqrt{1 + F^2}.$$

We can now express A, B, C, D in terms of F, using the abbreviations (5.19):

(5.24) $A = \sqrt{1 + F^2}$, $B = cE = cF$, $C = F/c$, $D = A = \sqrt{1 + F^2}$.

When these are inserted into (5.10), the transformation becomes

(5.25)
$$\bar{x} = \sqrt{1 + F^2}\,x + cFt$$
$$\bar{t} = (F/c)x + \sqrt{1 + F^2}\,t.$$

We now relate the parameter F to the speed v of the train as observed by the station master. The motorman perceives himself to be at rest with respect to his train. In his coordinate system, his position is described by

$$\bar{x} = 0.$$

In terms of the station master's coordinate system, see relations (5.25), $\bar{x} = 0$ reads

(5.26) $$\sqrt{1 + F^2}\,x + cFt = 0$$

and yields an expression for x in terms of t:

$$x = -\frac{cF}{\sqrt{1 + F^2}}\,t.$$

Thus from the station master's point of view, the motorman and his train are moving with velocity

$$v = -\frac{cF}{\sqrt{1+F^2}}.$$

When we solve this for F we obtain

$$F = \frac{-v}{\sqrt{c^2-v^2}}; \quad \text{also } \sqrt{1+F^2} = \frac{c}{\sqrt{c^2-v^2}}.$$

This enables us to write the transformation equations (5.25) in terms of the velocity as

$$\bar{x} = \frac{c}{\sqrt{c^2-v^2}} x - \frac{cv}{\sqrt{c^2-v^2}} t$$

$$\bar{t} = \frac{-v/c}{\sqrt{c^2-v^2}} x + \frac{c}{\sqrt{c^2-v^2}} t.$$

The quantity $c/\sqrt{c^2-v^2}$ occurs repeatedly in these and other equations. We shall therefore abbreviate this expression as g:

(5.27) $$g(v) = \frac{c}{\sqrt{c^2-v^2}} = \frac{1}{\sqrt{1-(v^2/c^2)}}.$$

Now the equations above relating $\bar{x}, \bar{t}$ to x, t can be written as

(5.28)
$$\bar{x} = g(v)(x - vt)$$

$$\bar{t} = g(v)\left(-\frac{v}{c^2}x + t\right)$$

The physical quantities length $\bar{x}$ and time $\bar{t}$ are real numbers. Therefore the function $g(v)$ in (5.28) must be real-valued. The definition (5.27) describes g in terms of a square root; for g to be real, the quantity under the square root must be nonnegative: $1 - v^2/c^2 \geq 0$, and this is the case if and only if

(5.29) $$v \leq c.$$

This shows that no observer may move with speed faster than light relative to any other observer. This sounds paradoxical since we can easily imagine three observers S, M, N, M moving with speed $\frac{2}{3}c$ relative to S and N moving with speed $\frac{2}{3}c$ relative to M, so that it would seem, at first blush, that N would move with speed $\frac{2}{3}c + \frac{2}{3}c = \frac{4}{3}c$ relative to S, i.e. with a speed exceeding c. Why this cannot be will be explained in detail in Chapter 6.

5.5 Rods Contract and Clocks Slow Down

The equations derived in the previous section,

(5.28)
$$\bar{x} = g(v)(x - vt)$$
$$\bar{t} = g(v)\left(-\frac{v}{c^2}x + t\right)$$

are the basis of Einstein's Special Theory of Relativity. We may or may not be disposed to accept them. But whether or not we like it, the fact remains that these are *necessarily* consequences of the invariance of the velocity of light.

In the remaining part of the book we shall consider some major consequences of these equations, consequences that shatter our complacency. To accept the basis of the Special Theory of Relativity without accepting its consequences is illogical. If we are willing to accept the evidence of the Michelson-Morley experiment, we should treat its logical consequences.

First we shall show how to use equations (5.28) to compare the lengths of objects as recorded by two different observers, for example the length of the train as it appears to the station master and the motorman, respectively. To do this, we imagine that as soon as the end of the train passes the station master, he sets off a flare. To him the flare up occurred at $x = 0$ and at the elapse of the time t it took for the whole train to go by, i.e. at $t = L/|v|$, where L is the length of the train and $|v|$ its speed, as observed by the station master. Setting $x = 0$, $t = L/|v|$ into (5.28) gives

$$\bar{x} = -g(v)v\frac{L}{|v|} = \pm g(v)L,$$

the minus sign holding if $v > 0$, the plus sign when $v < 0$.

Now $\bar{x}$ is the directed distance of the end of the train from the motorman and $|\bar{x}|$ is *the length of the train as it appears to the motorman*;

we denote this by $\overline{L}$. Then our last equation yields

(5.30) $$\overline{L} = g(v)L.$$

It is evident from formula (5.27) that for $v \neq 0$, g is greater than 1; therefore we conclude from (5.30) that for $v \neq 0$, $L < \overline{L}$.

In words: To the station master the length of the train appears shorter than it does to the motorman. This apparent shortening of objects which are in motion relative to the observer is called the *Fitzgerald-Lorentz contraction*. It takes place only in the direction of the motion.

This contraction seemed paradoxical to pre-Einstein physicists; dynamic explanations were sought, but none could be experimentally verified.

We turn now to an even more paradoxical and unforeseen consequence of the transformation laws (5.28). Let us denote by T the time $L/|v|$ shown on the station master's clock at the time he set off the flare. His coordinates for this event are $x = 0$, $t = T$. Setting these values into (5.28) shows that for the motorman this event occurred at time

(5.31) $$\overline{T} = gT.$$

As we saw earlier, g is > 1 if $v \neq 0$, so we conclude that $T < \overline{T}$.

Let us put it in words: the time it takes for the train to pass the station master S appears shorter to S than to the motorman. To understand this, we have to pin down what is special in the station master's relation to the process—the passage of the train—whose duration he is judging. What is special is that the events signalling the beginning and end of the process—the appearance of the locomotive, and the passage of the caboose[†]—take place at the spot where S is standing.

We can now state this principle in more general terms: Given any physical process, an observer S is called *special* if he is at rest relative to this process. The duration of a process as observed by a special observer is called the *proper time* of the process. To any other observer moving with velocity v relative to the process (and to the special observer), its duration appears greater than its proper time by the factor $g(v)$:

$$\text{moving clock time} = g(v) \cdot (\text{proper time})$$

or

(5.32) $$\text{proper time} = \frac{1}{g(v)} (\text{moving clock time}).$$

[†] This is what the last car of a train was called in those quaint old days when there were trains.

WHAT IS TIME ?

These considerations, consequences of the transformation equations, are essential for the understanding of all situations where there is relative motion at speeds near the speed of light. Let us consider space travel at such high speeds.

Suppose M is an astronaut heading straight for a distant star, 200 light years away from the earth. This means that earthbound astronomers have determined that it would take a light signal 200 years to traverse the distance to that star. Now suppose that M is hurtling through space with velocity $v = .995c$. The people on earth calculate that it will take him $200 \, c/v = 200/.995 \approx 201$ years to make the trip. Had M set out at age 25, his body would be 226 years old when he reached his destination. Surely he would have arrived a corpse. No! The crucial point is that 201 years is the duration of the flight in the experience of an earthbound observer S and his descendants, the people who stay home. M, the man who goes, is at rest relative to his space capsule and lives his experience in his proper time, recorded by the watch in his capsule. As far as the astronaut is concerned, the earth is moving away with velocity $-v$. Therefore time, as measured on earth, is related to M's time, the proper time, by (5.32):

$$\text{Astronaut's proper time} \approx \frac{1}{g(v)}(201) \approx \sqrt{1 - .99}\,(201) = 20.1 \text{ yrs.}$$

M will be about 45, not 226 years old when he reaches the star.

That the faster an astronaut travels the more slowly he ages gives us hope of men living long enough to visit the stars, all way out beyond the solar system. Yet you may be disposed to retort: Such subtle arguments are good, clean fun, but would any hard-headed astronaut be prepared to set out on a 200 year journey because it had been argued by a few long-haired professors that he would be only 20 years older when he got there? Not very likely. If I tear a page off my desk calendar and call today the first of September instead of the first of August, it doesn't make me any older physically. Next you will be telling me that if I forget to wind my watch, then I'll stop aging when it stops—and live forever!

No, the point is that *each* physical phenomenon runs its natural course in the system in which it rests, and *life is a physical phenomenon*. The moving astronaut lives his regular life in his space capsule. He does not have any benefit from the fact that an observer on a different system (which moves with very high speed relative to him) thinks that he lives very much longer. At this moment there are many galaxies which move relative to the earth with fantastic speed, nearly the velocity of light. If there were in such a galaxy a star with intelligent observers, they would think that we humans are practically immortal. This does not do us much good. However, for some purposes this difference in aging is of great use; while according to our system of accounting, an astronaut might need 200 years to reach a distant object, in his time scale he would need only 20 years and thus be able to survive his trip.

You may be quite bewildered and upset by our argument. But remember that your experience in life has been in systems of very slow motion, and there is nothing which could prepare your imagination to experiences of high speed travel in outer space. Whenever experience fails us, insight and intuition will fail likewise. Our only guide is our reason strengthened by mathematical argument. We might be wrong in our extrapolations, but until now the predictions of science have been more frequently verified than falsified.

Do you know what a radioactive substance is? It consists of a large number of atoms, of which, during a given period, a certain percentage disintegrates or dies. Uranium atoms, when placed in a cyclotron, are made to travel at nearly the velocity of light—just as we suppose M to do. It is found that uranium subjected to such cyclotron experience decays much more slowly than uranium subjected to ordinary terrestial experience. Here is evidence in favor of Einstein's time contraction formula. And isn't our aging a physiological process whose rate is that at which tissue and that sort of thing decay?

Other important evidence for the contraction of time intervals in moving systems comes from cosmic radiation, consisting of heavy particles from outer space showering down upon the earth with very high speed. On entering our atmosphere, they collide with the particles in the air and smash their atoms into many subatomic splinters. Some of them are so unstable that they fall apart in a very short time. Among them are the muons, particles of a mass between that of the electron and that of the proton. They decompose so fast that, after $2 \cdot 10^{-6}$ seconds, half of them have disappeared. Now, muons are created where our atmosphere begins, about $3 \cdot 10^6$ cm above ground. Even if they travelled with the speed of light $c = 3 \cdot 10^{10}$ cm/sec, they would need the time

$$T = \frac{3 \cdot 10^6}{3 \cdot 10^{10}} \sec = 10^{-4} \text{ seconds}$$

to reach sea level. Since this travel time is about hundred times larger than their half-life, almost all of them would have disappeared. However, a large number of muons can be observed in the laboratory at ground level. The explanation is that the muons live and die by their own clock, their proper time. This is much slower than the clock of the physicist on the ground. One can calculate the survival chance of a muon in its high speed frame and finds satisfactory agreement with observation.

As an exercise let us calculate how fast a muon would have to travel in order that the travel time $\bar{t} = 10^{-4}$ sec. measured on earth reduce to its half life, $t = 2 \cdot 10^{-6}$ sec. By (5.31), we have $t = \bar{t}/g(v) = \bar{t}\sqrt{1 - v^2/c^2}$, where v is the desired speed of the muon relative to the earth. An easy calculation yields $v = c\sqrt{1 - 4 \cdot 10^{-4}}$, i.e., $v = 0.9998c$. This high speed agrees with other considerations.

CHAPTER SIX

Relativistic Addition of Velocities

In deriving Einstein's transformation equations (5.28) in Section 5.4, we observed that they contain the quantity $g(v) = [1 - (v^2/c^2)]^{-1/2}$; and that therefore they only make sense if the relative speed v does not exceed c, the speed of light. So it is implicit in Einstein's theory—made hereby explicit—that no relative speed can exceed the speed of light.

This consequence jolts our intuition; we might have hoped, for example, that from a rocket going at nearly the velocity of light, another rocket could be launched at high velocity relative to the first, and so on until one obtains a velocity which, relative to an observer on earth, exceeds the speed of light. In this chapter we explain why this cannot be by showing that the velocities of platforms moving relative to one another are combined *not* by ordinary addition, but according to Einstein's famous law of relativistic addition.

6.1 Einstein's Law of Relativistic Addition

In Chapter 5 we have shown that the space-time coordinates x, t of an event observed by S and the space-time coordinates $\bar{x}, \bar{t}$ of that same event observed by M are related to each other by equations (5.28). The quantity v occurring in these equations is the speed with which M appears to be moving, as observed by S. The space-time coordinates $\hat{x}, \hat{t}$ of yet another observer, N, are similarly related to the space-time coordinates $\bar{x}, \bar{t}$ of M. The question we raise is:

How are the coordinates $\hat{x}, \hat{t}$ related to x, t? We can obtain the answer by substituting the expressions for $\bar{x}, \bar{t}$ in terms of x, t into the equations for $\hat{x}, \hat{t}$ in terms of $\bar{x}, \bar{t}$. These substitutions are most easily performed by first rewriting Einstein's transformation law in the language of vectors and matrices, as explained in Chapter 4.

We introduce the space-time vectors $\begin{pmatrix} x \\ t \end{pmatrix}$ and $\begin{pmatrix} \bar{x} \\ \bar{t} \end{pmatrix}$, and write equations (5.28) in the form

$$\begin{pmatrix} \bar{x} \\ \bar{t} \end{pmatrix} = g(v) \begin{pmatrix} 1 & -v \\ -v/c^2 & 1 \end{pmatrix} \begin{pmatrix} x \\ t \end{pmatrix}, \quad \text{where } g(v) = \frac{1}{\sqrt{1 - v^2/c^2}}.$$

We abbreviate by $L(v)$ the matrix appearing on the right:

$$(6.1) \qquad L(v) = g(v) \begin{pmatrix} 1 & -v \\ -v/c^2 & 1 \end{pmatrix}.$$

$L(v)$ is called a *Lorentz matrix*, corresponding to the velocity v. Einstein's transformation equations can then be simply written as

$$(6.2) \qquad \begin{pmatrix} \bar{x} \\ \bar{t} \end{pmatrix} = L(v) \begin{pmatrix} x \\ t \end{pmatrix}.$$

In words: The space-time vector $\begin{pmatrix} \bar{x} \\ \bar{t} \end{pmatrix}$ of observer M is obtained from the space-time vector $\begin{pmatrix} x \\ t \end{pmatrix}$ of observer S through multiplication by the Lorentz matrix $L(v)$, where v is the velocity of M relative to S.

We note that when $v = 0$,

$$L(v) = L(0) = g(0) \begin{pmatrix} 1 & 0 \\ 0 & 1 \end{pmatrix} = 1I = I,$$

so that $\begin{pmatrix} \bar{x} \\ \bar{t} \end{pmatrix} = \begin{pmatrix} x \\ t \end{pmatrix}$, as it ought when M is not moving relative to S. The space-time vector $\begin{pmatrix} \hat{x} \\ \hat{t} \end{pmatrix}$ of N can be expressed similarly in terms of $\begin{pmatrix} \bar{x} \\ \bar{t} \end{pmatrix}$:

$$(6.3) \qquad \begin{pmatrix} \hat{x} \\ \hat{t} \end{pmatrix} = L(u) \begin{pmatrix} \bar{x} \\ \bar{t} \end{pmatrix},$$

where u is the velocity of observer N relative to M.

To write $\begin{pmatrix} \hat{x} \\ \hat{t} \end{pmatrix}$ in terms of $\begin{pmatrix} x \\ t \end{pmatrix}$, let us express $\begin{pmatrix} \bar{x} \\ \bar{t} \end{pmatrix}$ on the right in (6.3) by (6.2); we obtain

$$(6.4) \qquad \begin{pmatrix} \hat{x} \\ \hat{t} \end{pmatrix} = (L(u)L(v)) \begin{pmatrix} x \\ t \end{pmatrix},$$

where $L(u)L(v)$ is the matrix product of the two matrices $L(u)$ and

$L(v)$. Using the rules of matrix multiplication, we obtain

$$L(u)L(v) = g(u)\begin{pmatrix} 1 & -u \\ -u/c^2 & 1 \end{pmatrix} g(v)\begin{pmatrix} 1 & -v \\ -v/c^2 & 1 \end{pmatrix}$$

(6.5)
$$= g(u)g(v)\begin{pmatrix} 1 + \dfrac{uv}{c^2} & -u - v \\ -\dfrac{u+v}{c^2} & 1 + \dfrac{uv}{c^2} \end{pmatrix}$$

$$= g(u)g(v)\left(1 + \dfrac{uv}{c^2}\right)\begin{pmatrix} 1 & -\dfrac{u+v}{1 + \dfrac{uv}{c^2}} \\ -\dfrac{u+v}{1 + \dfrac{uv}{c^2}} \dfrac{1}{c^2} & 1 \end{pmatrix}.$$

With the abbreviation

(6.6)
$$\frac{u+v}{1 + \dfrac{uv}{c^2}} = w,$$

the matrix on the right is

$$\begin{pmatrix} 1 & -w \\ -w/c^2 & 1 \end{pmatrix}.$$

It is easy to show, and is left to the reader as exercise, that

$$g(u)g(v)\left(1 + \frac{uv}{c^2}\right) = \frac{1}{\sqrt{1 - (w^2/c^2)}} = g(w).$$

So (6.5) can be written in the elegant form

(6.7) $$L(u)L(v) = L(w),$$

where w is related to u, v by formula (6.6).

In words: *the product of two Lorentz matrices is a Lorentz matrix.*

We ask: does $L(v)$ have an inverse? From (6.6) we see that if $u = -v$, then $w = 0$; but we already know that $L(0) = I$, while (6.7), with

$u = -v$, tells us that

$$L(-v)L(v) = I,$$

so $L(v)$ has an inverse and this is $L^{-1}(v) = L(-v)$.

A collection of matrices having these multiplicative properties is called a *group* of matrices. Relation (6.4) can be written as

(6.8) $$\begin{pmatrix} \hat{x} \\ \hat{t} \end{pmatrix} = L(w) \begin{pmatrix} x \\ t \end{pmatrix}$$

and says that the space-time vector of observer N is obtained from the space-time vector of S through multiplication by $L(w)$. It follows then according to the matrix form of the Einstein transformation law that w is the velocity of N relative to S. Thus we have Einstein's famous *relativistic addition law for velocities* (6.6):

$$w = \frac{u + v}{1 + uv/c^2}.$$

We now show that if the velocities u and v do not exceed c, neither does their relativistic "sum" velocity w. For if $u \le c$ and $v \le c$, then $0 \le u^2/c^2 \le 1$ and $0 \le v^2/c^2 \le 1$; and so $0 \le 1 - u^2/c^2 \le 1$ and $0 \le 1 - v^2/c^2 \le 1$. Consequently their product is nonnegative:

$$\left(1 - \frac{u^2}{c^2}\right)\left(1 - \frac{v^2}{c^2}\right) = 1 - \frac{u^2 + v^2}{c^2} + \frac{u^2v^2}{c^4} \ge 0.$$

It follows that

$$\frac{u^2 + v^2}{c^2} \le 1 + \frac{u^2v^2}{c^2 c^2} \quad \text{and} \quad u^2 + v^2 \le c^2 + \frac{u^2v^2}{c^2}.$$

Now let us add $2uv$ to both sides of the last inequality:

$$u^2 + 2uv + v^2 = (u+v)^2 \le c^2 + 2uv + \frac{u^2v^2}{c^2} = c^2\left(1 + \frac{u}{c}\frac{v}{c}\right)^2$$

Dividing by $(1 + \frac{u}{c}\frac{v}{c})^2$ yields

$$w^2 = \left(\frac{u+v}{1 + \frac{uv}{c^2}}\right)^2 \le c^2, \quad \text{and hence} \quad w \le c.$$

6.2 Rescaling Velocities

In Chapter 4, we showed how to describe rotations with the aid of orthogonal matrices. In the present section, we shall see how similar this is to describing Einstein's transformations with the aid of Lorentz matrices. By exploiting this analogy we shall defined a parameter V which will play a role analogous to that played by the angle of rotation in direct isometries.

In studying such isometries in Sections 4.12–13, we required that for every vector $\binom{x}{y}$, the transformed vector

$$\begin{pmatrix}\bar{x}\\\bar{y}\end{pmatrix} = \begin{pmatrix}a & b\\c & d\end{pmatrix}\begin{pmatrix}x\\y\end{pmatrix}$$

have the same length as the original:

$$\bar{x}^2 + \bar{y}^2 = x^2 + y^2.$$

We saw that all transformations by means of matrices of form

(6.9) $$R = \begin{pmatrix}a & -b\\b & a\end{pmatrix}$$

have this property, provided that

(6.10) $$a^2 + b^2 = 1.$$

Using the identity

(6.10)' $$\cos^2\theta + \sin^2\theta = 1$$

we were able to parametrize all solutions of (6.10) by setting

(6.11) $$a = \cos\theta, \quad b = \sin\theta.$$

Substituting this into the matrix (6.9) led to

(6.12) $$R(\theta) = \begin{pmatrix}\cos\theta & -\sin\theta\\\sin\theta & \cos\theta\end{pmatrix}.$$

R has the geometric interpretation of rotation through an angle θ. Since rotation through ϕ followed by rotation through θ has the same effect as

RELATIVISTIC VELOCITIES

rotation through $\theta + \phi$, it follows that

(6.13) $$R(\theta)R(\phi) = R(\theta + \phi).$$

On the other hand, if we carry out the matrix multiplication $R(\theta)R(\phi)$, we see that (6.13) is equivalent with the addition formulas

(6.14) $$\cos(\theta + \phi) = \cos\theta\cos\phi - \sin\theta\sin\phi$$
$$\sin(\theta + \phi) = \cos\theta\sin\phi + \sin\theta\cos\phi.$$

In Section 5.4, see equations (5.18), we similarly studied matrices with the following analogous property:
For every vector $\begin{pmatrix} x \\ y \end{pmatrix}$, we require that the transformed vector

$$\begin{pmatrix} \bar{x} \\ \bar{y} \end{pmatrix} = \begin{pmatrix} A & E \\ F & D \end{pmatrix} \begin{pmatrix} x \\ y \end{pmatrix}$$

satisfy

$$\bar{x}^2 - \bar{y}^2 = x^2 - y^2.$$

We saw that all transformations by means of matrices of form

(6.15) $$\begin{pmatrix} A & -G \\ -G & A \end{pmatrix}$$

have this property, provided that

(6.16) $$A^2 - G^2 = 1.$$

[$-G$ is the quantity we called F in Section 5.4.]
We are looking for a parametrization of solutions of (6.16) analogous to (6.11). We claim that it is given by the *hyperbolic functions*

(6.17) $\quad A = \cosh V = \tfrac{1}{2}(e^V + e^{-V}), \qquad G = \sinh V = \tfrac{1}{2}(e^V - e^{-V}).$

The exponential function e^x appearing in this definition is familiar to us from Chapter 2 on Growth Functions (see Section 2.1). We recall the functional equation (2.22) from p. 42, $f(a + b) = f(a) \cdot f(b)$; for $f(x) = e^x$, we have

$$e^{a+b} = e^a e^b,$$

a sort of addition formula. With its help it is easy to verify these immediate consequences of definitions (6.17):

(6.18) $$\cosh^2 V - \sinh^2 V = 1$$

and

(6.19)
$$\cosh(U+V) = \cosh U \cosh V + \sinh U \sinh V$$
$$\sinh(U+V) = \cosh U \sinh V + \sinh U \cosh V.$$

We leave the details to the reader.

The similarity between the trigonometric identities (6.10)′, (6.14) and their hyperbolic analogues (6.18), (6.19) leads us to suspect that the latter can be deduced from the former. This can indeed be done in these steps:

1. Represent the exponential and trigonometric functions by their power series

$$e^x = \sum_{n=0}^{\infty} \frac{x^n}{n!}, \quad \cos x = \sum_{n=0}^{\infty} (-1)^n \frac{x^{2n}}{(2n)!}, \quad \sin x = \sum_{n=0}^{\infty} (-1)^n \frac{x^{2n+1}}{(2n+1)!},$$

extend their domains to complex values of x, and verify that (6.10)′, (6.14) are valid for complex arguments.

2. Use these series representations to verify Euler's identity

(E) $$e^{i\theta} = \cos\theta + i\sin\theta, \quad i = \sqrt{-1},$$

and use (E) and (6.17) to establish the relations

(E′) $$\cosh V = \cos iV, \quad \sinh V = -i \sin iV$$

between the hyperbolic[†] and trigonometric functions.

3. Use (E′) to show the equivalence of (6.10)′ with (6.18), and of (6.14) with (6.19).

Equation (6.18) shows that A and G defined in (6.17) do indeed satisfy (6.16). Therefore we set (6.17) into the matrix (6.15). We obtain

(6.20) $$H(V) = \begin{pmatrix} \cosh V & -\sinh V \\ -\sinh V & \cosh V \end{pmatrix}.$$

[†] The name "hyperbolic functions" indicates a geometric relation of $\cosh\theta$, $\sinh\theta$ to the hyperbola $x^2 - y^2 = 1$ analogous to the relation that the circular functions $\cos\theta$, $\sin\theta$ bear to the circle $x^2 + y^2 = 1$.

RELATIVISTIC VELOCITIES

We can easily show that the matrix H satisfies the functional equation

(6.21) $$H(U + V) = H(U)H(V)$$

by carrying out the matrix multiplication and using the addition formulas (6.19).

We now switch back from the variables $y, \bar{y}$ to $t, \bar{t}$ by putting $y = ct$, $\bar{y} = c\bar{t}$ into the transformation equations with matrix (6.15). We obtain the mapping from $\begin{pmatrix} x \\ t \end{pmatrix}$ to $\begin{pmatrix} \bar{x} \\ \bar{t} \end{pmatrix}$

$$\begin{pmatrix} \bar{x} \\ \bar{t} \end{pmatrix} = \begin{pmatrix} A & -Gc \\ -G/c & A \end{pmatrix} \begin{pmatrix} x \\ t \end{pmatrix} = L \begin{pmatrix} x \\ t \end{pmatrix}.$$

We replace A and G from (6.17); then

(6.22) $$L(V) = \begin{pmatrix} \cosh V & -c\sinh V \\ -\dfrac{\sinh V}{c} & \cosh V \end{pmatrix}$$

gives the desired parametrization of Lorentz matrices in terms of the parameter V. The product of two such matrices yields, thanks to the addition formulas (6.19), the relation

(6.21)′ $$L(U)L(V) = L(U + V).$$

How is this related to the parametrization (6.1) of $L(v)$ in terms of the velocity v, given in the previous section? We equate corresponding elements of (6.1) and (6.22):

$$\cosh V = g(v), \qquad c\sinh V = vg(v).$$

From this we deduce that

$$\frac{\sinh V}{\cosh V} = \frac{v}{c}.$$

This ratio of the hyperbolic sine to the hyperbolic cosine is called *hyperbolic tangent* of V; by the definitions (6.17), we find

$$\frac{e^V - e^{-V}}{e^V + e^{-V}} = \frac{v}{c}.$$

We multiply top and bottom of the left side by e^V, obtaining

$$\frac{e^{2V} - 1}{e^{2V} + 1} = \frac{v}{c},$$

and we solve this first for e^{2V}, then for V:

$$e^{2V} = \frac{1 + \dfrac{v}{c}}{1 - \dfrac{v}{c}}.$$

Taking the natural logarithms of each side, we get

$$2V = \ln\frac{1 + \dfrac{v}{c}}{1 - \dfrac{v}{c}},$$

and

(6.23) $$V = \frac{1}{2}\ln\frac{1 + \dfrac{v}{c}}{1 - \dfrac{v}{c}}.$$

This function V of v is called the *rescaled velocity*. Observe that as v approaches c, V increases to ∞; also when $v = 0$, $V = 0$.

In Section 6.1 we have shown that the product $L(u) \cdot L(v)$ of two Lorentz matrices is a Lorentz matrix $L(w)$, and that the velocity parameter w is related to the parameters u and v of the factors by Einstein's law (6.6) for the relativistic addition of velocities. The corresponding product $L(U)L(V)$ of matrices of form (6.22) obey relation (6.21)' which shows that the rescaled velocity W in the product is related much more simply to the rescaled velocities U and V of the factors; the law of combination is ordinary addition:

(6.24) $$W = U + V.$$

Exercise 6.1. Using the definitions $\cosh V = \frac{1}{2}(e^V + e^{-V})$, $\sinh V = \frac{1}{2}(e^V - e^{-V})$, verify the identities (6.18) and (6.19).

Exercise 6.2. The isometry

$$R(\theta) = \begin{pmatrix} \cos\theta & -\sin\theta \\ \sin\theta & \cos\theta \end{pmatrix}$$

RELATIVISTIC VELOCITIES

can be written in the equivalent form

$$R(\theta) = \cos\theta \begin{pmatrix} 1 & -\tan\theta \\ \tan\theta & 1 \end{pmatrix}.$$

(a) Show that (6.22) can be written in the analogous form

$$L(V) = \cosh V \begin{pmatrix} 1 & -c\tanh V \\ -\dfrac{\tanh V}{c} & 1 \end{pmatrix}$$

where $\tan V = \sinh V / \cosh V$.

(b) Use this form to find the matrix product $L(U)L(V)$ and thus derive the law of addition for the hyperbolic tangent

(6.25) $$\tanh(U + V) = \frac{\tanh U + \tanh V}{1 + \tanh U \tanh V}.$$

(c) Derive the same addition law from (6.19), and express the result also in terms of exponentials.

Exercise 6.3. Using Einstein's law $\dfrac{u + v}{1 - (uv/c^2)} = w$ and relation (6.23) to rescale velocities, verify directly that $U + V = W$.

6.3 Experimental Verification of Einstein's Law

Einstein's law of relativistic addition of velocities, expressed in equation (6.6) as

(6.26) $$w_{\text{rel}} = \frac{u + v}{1 + uv/c^2}$$

is so contrary to our everyday experience with combined velocities

(6.26)′ $$w_{\text{classic}} = u + v,$$

that an experimental verification is highly desirable. When u and v are both small compared to c, as in our daily experience with motions of bodies on earth, the difference between (6.26) and (6.26)′ is so small that it is very hard to detect experimentally. But when both velocities u and v are comparable to c, then the relativistic formula (6.26) gives a vastly different answer from the classical case. It is a surprising mathematical fact that when only one of the velocities, say u, is comparable in size to c, and the other velocity v is small compared to c, the relativistic formula (6.26) yields a strikingly different answer from the classical one. This is easily seen by using a little calculus; but first it is helpful to introduce as new variables the ratios of u, v to c:

(6.27) $$r = \frac{u}{c}, \qquad s = \frac{v}{c},$$

and to rewrite (6.26) as

(6.28) $$w_{\text{rel}} = c\frac{r + s}{1 + rs}.$$

We assume that s is small. According to calculus, a differentiable function $f(s)$ is well approximated for small s by the linear function of s

(6.29) $$f(s) \approx f(0) + f'(0)s,$$

where $f'(0)$ is the derivative of f at $s = 0$, and the error in this approximation is less than some constant multiple of s^2, called $0(s^2)$. We now take

$$f(s) = \frac{r + s}{1 + rs};$$

then

$$f'(0) = \left.\frac{df}{ds}\right|_{s=0} = 1 - r^2,$$

so from (6.29) we get

$$\frac{r + s}{1 + rs} = r + (1 - r^2)s + 0(s^2).$$

Setting this into formula (6.28) and reverting to the original variables, we find that

(6.30) $$w_{\text{rel}} = u + (1 - r^2)v + 0(v^2/c)$$
$$\approx u + \left(1 - \frac{u^2}{c^2}\right)v.$$

When u/c is not small, then (6.30) is appreciably different from (6.26).

The following physical set-up involves a large u/c and a tiny v/c: Light propagates in water with speed $u \approx .75c$. Suppose water is placed in a tube, and the tube is set in motion with velocity v in the laboratory; then the light in the moving tube appears to an observer who is stationary in the laboratory to be propagating with velocity

$$(6.31) \qquad w_{\text{rel}} = u + \left(1 - \frac{u^2}{c^2}\right)v.$$

The experiment described above was actually performed by the French physicist Fizeau; his measured value was in good agreement with (6.31), constituting a striking verification of Einstein's formula.

But Fizeau's experiment pre-dated Einstein's theory by about half a century! The experiment had been suggested early in the 19th Century by Fresnel to test his theory for the propagation of light in a mysterious "luminiferous" medium called *ether* which is at absolute rest, while "ponderable" matter moves through it like knife through butter. Fresnel announced his formula in 1818, and Fizeau's experiment was regarded as a striking verification of the properties of ether! Einstein's theory dispenses with the notions of absolute rest and ether. It is ironic that the very experiment which, in the middle of the 19th Century, was regarded as one of the finest proofs of the existence of ether turned out, in the 20th Century, to be one of the experimental verifications of Einstein's diametrically opposite theory of relativity.

6.4 Rescaled Velocity Revisited

In Section 6.2 we have shown how to rescale velocity v by measuring it in terms of an appropriately chosen function $V(v)$ of velocity; this rescaling function

$$V = \frac{1}{2} \ln \frac{1 + v/c}{1 - v/c}$$

was given in equation (6.23). The virtue of rescaling is that it translates Einstein's somewhat complicated relativistic addition (here denoted by $\oplus$) of velocities u and v,

$$w = \frac{u + v}{1 + uv/c^2} = u \oplus v$$

defined by (6.6), into ordinary addition:

$$W = U + V.$$

We arrived at the rescaling by observing a fortuitous analogy between Lorentz and rotation matrices.

In this section, we show that there is nothing fortuitous about it by basing the possibility of rescaling on an important *algebraic* property of Einstein's law of addition $\oplus$, namely its associativity:

$$(6.32) \qquad u \oplus (v \oplus y) = (u \oplus v) \oplus y.$$

Exercise 6.4. Using formula (6.23), verify that the operation $\oplus$ is associative.

If you have done the exercise, you will appreciate the argument without any calculations we are about to give for the associative property.

Consider observers A, B, C, D moving relative to one another so that

u is B's velocity relative to A,
v is C's velocity relative to B,
y is D's velocity relative to C;

then

C's velocity relative to A is $u \oplus v$, and
D's velocity relative to B is $v \oplus y$.

The velocity of D relative to A can be reckoned in two ways: using observer B, or using observer C. The first way gives $u \oplus (v \oplus y)$, the second $(u \oplus v) \oplus y$. Their equality is the associative property (6.32).

We now formulate and prove a very general proposition about associative binary operations. A *binary* operation on numbers is a function which assigns a single number to every pair of numbers u and v: $F(u, v) = w$.

Theorem. Let $F(u, v)$ be a binary operation which, like addition, is associative:

$$(6.33) \qquad F(u, F(v, y)) = F(F(u, v), y) \qquad \text{for all } u, v, y,$$

and which, like addition, satisfies

$$(6.34) \qquad F(u, 0) = u.$$

Then it is possible to rescale all the numbers, i.e. it is possible to find a function h and set

$$U = h(u) \qquad \text{for all } u$$

so that the binary operation F is turned into addition:

(6.35) $$h(F(u,v)) = h(u) + h(v) = U + V.$$

The task is to find h. Note that by setting $v = 0$ in (6.35) and using (6.34), we find that

$$h(0) = 0.$$

Proof. If there is a differentiable function h satisfying (6.35) for all u, v, then we may differentiate (6.35) with respect to v. We use the chain rule and denote the derivative of $F(u, v)$ with respect to its second argument by F_2'; we obtain

(6.36) $$h'(F(u,v))F_2'(u,v) = h'(v).$$

Setting $v = 0$ and using (6.34), we find that

$(6.36)_0$ $$h'(u)F_2'(u,0) = k$$

where k abbreviates $h'(0)$. Since $h(0) = 0$, we can now determine $h(u)$ by solving $(6.36)_0$ for $h'(u)$ and integrating:

(6.37) $$h(u) = k\int_0^u \frac{1}{F_2'(z,0)}\,dz.$$

We now verify that the function h constructed in (6.37) indeed satisfies (6.35) for all u, v. First note that if $v = 0$ and $F(u, v)$ is used as the argument of h in (6.37), then by (6.34)

$$h(F(u,0)) = h(u),$$

so that (6.35) holds for arbitrary u and for $v = 0$. To show that (6.35) holds also for arbitrary v, we first establish that its two sides have identical v-derivatives, i.e. that (6.36) holds for all u, v. From this, together with the validity of (6.35) at $v = 0$, we shall deduce that (6.35) holds for all u, v.

First we differentiate the associative rule (6.33) with respect to y and set $y = 0$:

$$F_2'(u, F(v,0))F_2'(v,0) = F_2'(F(u,v),0).$$

Since $F(v, 0) = v$, we have

$$F_2'(u,v)F_2'(v,0) = F_2'(F(u,v),0),$$

so

$$F_2'(u,v) = \frac{F_2'(F(u,v),0)}{F_2'(v,0)}.$$

When this is substituted into the left member of (6.36) it becomes

(6.38) $$\frac{h'(F(u,v))F_2'(F(u,v),0)}{F_2'(v,0)};$$

its numerator is precisely the left side of (6.36)$_0$ with u replaced by $F(u,v)$, and hence may be replaced by the right side of (6.36)$_0$; so (6.38) can be written

(6.39) $$\frac{k}{F_2'(v,0)}.$$

By (6.36)$_0$, $F_2'(v,0) = k/h'(v)$, so (6.39) becomes $h'(v)$. Hence with the function $h(u)$ defined by (6.37), we have the identity

$$\frac{d}{dv}h(F(u,v)) = h'(F(u,v))F_2'(u,v) = h'(v).$$

Thus $h(F(u,v)) = h(v) + A(u)$, where A may depend on u. Put $v = 0$ and use (6.34) to find $h(u) = h(0) + A(u) = A(u)$, and so (6.35) is proved and hence also the theorem.

We have not only proved that it is possible to rescale; we have actually constructed a function h, see formula (6.37), that does the rescaling. Observe that if the function h rescales so that (6.35) holds, then any constant multiple of h also accomplishes this.

Let us see what our formula for h gives when the binary operation $F(u,v)$ is relativistic addition (6.6), which, as we have seen, fulfills the hypotheses of our theorem. For

$$F(u,v) = \frac{u+v}{1+uv/c^2}$$

we get

$$F_2'(u,0) = \frac{d}{dv}F(u,v)\bigg|_{v=0} = \left(1 - \frac{u^2}{c^2}\right)\left(1 + \frac{uv}{c^2}\right)^{-2}\bigg|_{v=0} = 1 - \frac{u^2}{c^2}.$$

So formula (6.36) gives

$$h(u) = k \int_0^u \frac{dz}{1-(z/c)^2} = kc^2 \int_0^u \frac{dz}{c^2-z^2} = \frac{kc}{2} \int_0^u \left[\frac{1}{c+z} + \frac{1}{c-z} \right] dz$$

$$= \frac{kc}{2} \ln \frac{c+u}{c-u} = \frac{kc}{2} \ln \frac{1+u/c}{1-u/c}.$$

If we choose the constant k so that $h'(0) = k = 1/c$, then we have constructed the function h that yields our old rescaling (6.23).

The quantity $F_2'(u, 0)$ gives the "infinitesimal" binary operation $F(u, v) \approx u + F_2'(u, 0)v$, valid within an error $0(v^2)$ for very small values of v.

We have shown that an h can be determined from a mere knowledge of the function $F_2'(u, 0)$. Once we know h, we can reconstruct the whole binary operation using formula (6.35).

Exercise 6.5. Given

$$h(u) = \frac{1}{2} \ln \frac{1+u/c}{1-u/c},$$

use (6.36) to reconstruct the relativistic addition (6.6).

Exercise 6.6. (a) Show that

$$F(u, v) = u + v + uv$$

is an associative binary operation satisfying (6.34).
(b) Find the function h that linearizes this operation.
(c) Verify that h linearizes F.

We can express all this by saying: An associative binary operation can be reconstructed if the infinitesimal binary operation is known. We note, moreover, that $F_2'(u, 0)$ is precisely the coefficient of v in the linear approximation (6.30) of relativistic addition; this was the quantity Fresnel predicted theoretically and Fizeau measured experimentally.

The above derivation of the rule of addition of relativistic velocities from the associative property and knowledge of the sum when one of the velocities is small is a curious mixture of algebra and calculus. The same ingredients are combined and exploited systematically in a branch of mathematics called theory of Lie groups and algebras. This theory was created by the great Norwegian mathematician Sophus Lie in the nineteenth Century and came to full flower in the twentieth Century. Thus, someone in the nineteenth Century familiar with Lie's mathematics and Fresnel's formula (and Fizeau's verification of it) could have discovered the special theory of relativity! Think this over! You will understand why scientists call mathematics our sixth sense with which to experience reality.

CHAPTER SEVEN

Energy

In investigating laws of science, one has first of all to decide what depends on what. Does the dynamics of a material body depend on its color or temperature? Does it depend on the place or the time at which the experiment is performed and so on? If one has decided that the outcome is independent of some parameters, new and deep insights on the expected laws can be obtained. The entire theory of special relativity could be deduced from the experimental fact that the velocity of light is the same for all observers who are moving relative to one another with constant velocity, and the requirement that *the laws of mechanics should be the same for all such observers*.

Now, if we take the equations describing these laws for one observer and transform them into the equations for any other one, we will obtain formally new equations, depending upon the relative velocity of the two observers. But the equations must be the same, by our requirement. Thus, all admissible equations of the theory must be unchanged under such a transformation. This demand of invariance puts a great restriction on all possible equations of the theory.

Laws of invariance are an important guide if we enter an unknown field of knowledge. The simplest applications go back to classical antiquity. Indeed, the simplest transformation which occurs in science is reflection. When Archimedes studied the law of the lever, he used this concept, demanding symmetry between left and right.

When we ask for the gravitational field of a homogeneous heavy sphere, we are sure this field is radially symmetric, and this greatly simplifies our methods for finding it. In this section, we shall give a number of examples of how the reasoning works. Let us illustrate the method first by a famous historic example.

7.1 The Two Body Impact Problem in Classical Mechanics

The laws of mechanics which seem to us today so evident were developed through careful study by Galileo, Stevin, Descartes and Huygens and finally brought into a simple but powerful system of laws by Newton in his famous *Principia Mathematica*. We have already shown some of the ingenious thought processes of these pioneers. We shall now describe the arguments Huygens used to find the correct laws for the impact of two material bodies moving along a straight line in opposite directions.

Suppose a body B_1 with mass m_1 moving along the x-axis with velocity v_1 meets a body B_2 with mass m_2 travelling along the same straight line with velocity v_2. If B_1 and B_2 are perfectly elastic, no kinetic energy is lost on collision. How do they move after colliding?

Consider first the special symmetric case where

$$m_1 = m_2 \quad \text{and} \quad v_2 = -v_1.$$

If B_1 and B_2 are heading toward each other, they will clash and recoil; call the recoil velocities w_1 and w_2, respectively. For reasons of symmetry $w_2 = -w_1$. To determine the speed $|w_1| = |w_2|$ after the collision, we shall use the law of conservation of energy; we recall that for a given mass the energy of motion (kinetic energy) depends on the velocity of the body. Before the clash, B_1 has the energy $E_1 = \tfrac{1}{2}m_1 v_1^2$, and B_2 has the energy $E_2 = \tfrac{1}{2}m_2 v_2^2 = \tfrac{1}{2}m_1 v_1^2 = E_1$. Thus the total energy of the system before the collision is $E_1 + E_2 = m_1 v_1^2$. Conservation of energy demands that the total energy after the collision, namely $m_1 w_1^2$, be equal to the energy before the collision. We find that necessarily

$$w_1 = -v_1 = v_2 \quad \text{and} \quad w_2 = -v_2 = v_1.$$

Thus the solution of our special problem is: After the clash each body changes the direction of its motion but continues with the same speed. Figure 7.1 suggest a succinct reformulation: in a symmetric collision B_1 and B_2 interchange their velocities.

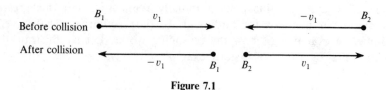

Figure 7.1

ENERGY

Now, let us relax our assumptions a little. We still keep the condition $m_1 = m_2$, but we allow $|v_1|$ and $|v_2|$ to be different. The symmetry seems lost. Now Huygens, in an interesting anticipation of Einstein's principle of relativity, restored symmetry. He did this by introducing an observer M, who moves with velocity u along the same line, and by asserting that the laws of mechanics valid for a stationary observer S must hold also for M. While relative to S, B_1 and B_2 move with velocities v_1 and v_2, relative to M they move with velocities $v_1 - u$ and $v_2 - u$. See Figure 7.2a. Now Huygens chooses

$$u = \frac{v_1 + v_2}{2}.$$

Then the velocities of B_1 and B_2 relative to the moving observers are

$$v_1 - u = \frac{v_1 - v_2}{2} \quad \text{and} \quad v_2 - u = \frac{v_2 - v_1}{2} = -(v_1 - u),$$

respectively. Relative to M, symmetry has been established. But for M the laws of mechanics are the same as for S, so M predicts that after collision the bodies will recoil with exchanged velocities. M will see B_1 recoil with velocity $v_2 - u$ and B_2 recoil with velocity $v_1 - u$. Since M moves with velocity u relative to S, we immediately deduce that after collision, S sees B_1 move with velocity

$$w_1 = (v_2 - u) + u = v_2$$

and B_2 with velocity

$$w_2 = (v_1 - u) + u = v_1.$$

See Figure 7.2.

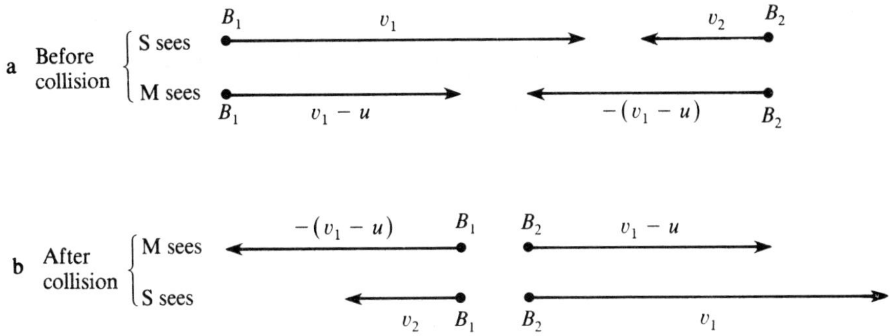

Figure 7.2

We have proved that after impact the bodies move with exchanged velocities. We have deduced this new insight from the principle of *invariance* of the laws of mechanics *under change of observers*. It is interesting that Huygens gave this argument about twenty years before Newton formulated the theory of mechanics in his famous three fundamental laws.

We shall now deal with the most general situation where the bodies B_1, B_2 have arbitrary masses m_1, m_2 and velocities v_1 and v_2, and determine the velocities w_1, w_2 after impact using only the law of conservation of energy and the principle of invariance.

The energy conservation law implies that

$$(7.1) \qquad \tfrac{1}{2}m_1 v_1^2 + \tfrac{1}{2}m_2 v_2^2 = \tfrac{1}{2}m_1 w_1^2 + \tfrac{1}{2}m_2 w_2^2.$$

Relative to an observer travelling with velocity u, B_1 is moving with velocity $v_1 - u$ before the clash and with velocity $w_1 - u$ after; analogously, B_2 moves with velocity $v_2 - u$ before and with $w_2 - u$ after. Relative to the observer, the energy conservation law is

$$(7.2) \qquad \tfrac{1}{2}m_1(v_1-u)^2 + \tfrac{1}{2}m_2(v_2-u)^2 = \tfrac{1}{2}m_1(w_1-u)^2 + \tfrac{1}{2}m_2(w_2-u)^2;$$

this holds for all u. Expanding the squares and using (7.1), we get

$$(7.3) \qquad m_1 v_1 + m_2 v_2 = m_1 w_1 + m_2 w_2.$$

Relation (7.3) is the law of conservation of momentum, derived from the energy conservation law and the principle of invariance under change of observer.

The system (7.1), (7.3) for w_1 and w_2 has two solutions since (7.1) is quadratic and (7.3) linear. One of these, $w_1 = v_1$, $w_2 = v_2$, is physically impossible and mathematically trivial. It is physically impossible because it leaves the velocities of the bodies unaltered by the alleged collision as if one had passed like a ghost through the other. It is mathematically obvious because we clearly conserve energy and momentum when we do not alter velocities.[†] To find the changed velocities after impact, we write (7.1) in the equivalent form

$$(7.4) \qquad m_1(v_1 - w_1)(v_1 + w_1) + m_2(v_2 - w_2)(v_2 + w_2) = 0.$$

[†]Mathematicians label obvious results "trivial". Readers interested in the origin of this adjective should look up the meanings of "trivium" and "quadrivium" in medieval education.

From (7.3) we find $m_2(v_2 - w_2) = -m_1(v_1 - w_1)$ and substitute this into (7.4), obtaining

$$m_1(v_1 - w_1)(v_1 + w_1) - m_1(v_1 - w_1)(v_2 + w_2) = 0.$$

Since for a nontrivial solution we can divide by $m_1(v_1 - w_1)$, we obtain the linear equation

(7.5) $$v_1 + w_1 = v_2 + w_2.$$

This together with (7.3) constitutes a pair of linear equations with the unique solution

$$w_1 = \frac{2m_2 v_2 + v_1(m_1 - m_2)}{m_1 + m_2}, \qquad w_2 = \frac{2m_1 v_1 + v_2(m_2 - m_1)}{m_1 + m_2}.$$

Note that in the case $m_1 = m_2$, we get our old solution again: after impact the bodies move with exchanged velocities.

7.2 The Two Body Impact Problem in the Theory of Relativity

In this section we shall investigate how the theory of relativity treats the elastic collision of two bodies. We need to revise not only our notions of length and time, but also our notions of mass and energy.

It turns out that, in the theory of relativity, the mass of a body perceived by an observer depends on the velocity with which the body moves relative to that observer. We define the *rest mass* of a body to be its mass measured when the body is at rest with respect to the observer who makes the measurement.

Suppose a body with rest mass m is moving with velocity v relative to an observer. Denote its energy with respect to that observer by $E(m, v)$. What kind of a function is $E(m, v)$?

Let B_1 and B_2 be two bodies with rest masses m_1 and m_2, both moving with velocity v relative to an observer. We assume that the composite body obtained by joining B_1 and B_2 has rest mass $m_1 + m_2$ and that its energy is the sum of the energies of B_1 and B_2:

$$E(m_1 + m_2, v) = E(m_1, v) + E(m_2, v).$$

Such a function is called *additive* with respect to the variable m. We further assume that E is a continuous function of m. It is a simple but important fact that a continuous additive function is linear; thus

(7.6) $$E(m, v) = m\Phi(v).$$

To determine the function $\Phi(v)$ of the single variable v, we consider again the collision of two bodies, this time in the relativistic case. Suppose two bodies with rest masses m_1 and m_2 and velocities v_1 and v_2 relative to an observer collide. Denote their velocities after collision by w_1, w_2. We assume that the collision is elastic, i.e. that total energy is conserved:

(7.7) $$m_1\Phi(v_1) + m_2\Phi(v_2) = m_1\Phi(w_1) + m_2\Phi(w_2).$$

It is useful to rescale velocities, by using (6.23) of Section 6.2:

(7.8) $$V = \frac{1}{2}\ln\frac{1 + v/c}{1 - v/c}.$$

We now express Φ as function of the rescaled velocity V:

$$\Phi(v) = \phi(V).$$

Then the energy may be written

(7.9) $$E = m\phi(V).$$

In terms of the rescaled velocities, the law of conservation of energy reads

(7.10) $$m_1\phi(V_1) + m_2\phi(V_2) = m_1\phi(W_1) + m_2\phi(W_2).$$

This law must hold with respect to any observer. Denote by U the rescaled velocity of an observer; then by equation (6.24) the rescaled velocities of the two particles before and after collision are $V_1 - U$, $V_2 - U$ and $W_1 - U$, $W_2 - U$. Now the conservation law reads

(7.11) $$m_1\phi(V_1 - U) + m_2\phi(V_2 - U) = m_1\phi(W_1 - U) + m_2\phi(W_2 - U).$$

This is the analogue of (7.2), and it holds for all U.

7.3 Admissible Energy Functions

We shall call a function ϕ an *admissible energy function* if, for arbitrarily given V_1, V_2, m_1, m_2, equation (7.11) has a unique non-trivial solution W_1, W_2 valid for all U. By non-trivial we mean that $(W_1, W_2) \neq$

(V_1, V_2). [Of course $(W_1, W_2) = (V_1, V_2)$ is always a solution but means that no collision took place.] An example of an admissible energy function is $\phi(V) = \tfrac{1}{2}V^2$, the Newtonian energy function, as discussed in Section 7.1. We shall determine all admissible energy functions and then single out the one that represents relativistic energy.

Observe that (7.11), required to hold for all U, constitutes infinitely many equations for W_1, W_2. Since in general just two equations determine W_1, W_2 uniquely, we note that admissibility places a severe restriction on the class of functions ϕ.

We shall extract three out of the infinitely many equations (7.11) by differentiating (7.11) with respect to U and then setting $U = 0$ in the original and in the derived equations.†

Thus we find

(7.12) $$m_1\phi(V_1) + m_2\phi(V_2) = m_1\phi(W_1) + m_2\phi(W_2)$$

(7.12′) $$m_1\phi'(V_1) + m_2\phi'(V_2) = m_1\phi'(W_1) + m_2\phi'(W_2)$$

(7.12″) $$m_1\phi''(V_1) + m_2\phi''(V_2) = m_1\phi''(W_1) + m_2\phi''(W_2).$$

We claim that these three equations are compatible if and only if ϕ satisfies a differential equation of the form

(7.13) $$a\phi''(V) + b\phi'(V) + k\phi(V) = d$$

where d and the coefficients a, b, k are constants. That they are compatible if ϕ satisfies (7.13) is clear, since then (7.12″) is an immediate consequence of the two preceding equations.

We shall now sketch a proof that admissible energy functions must necessarily satisfy an equation of form (7.13). We can then integrate explicitly this simple differential equation with constant coefficients and verify that the solution function $\phi(V)$ satisfies (7.11) and is therefore admissible. We shall carry out this program in the case that interests us in Section 7.5.

In order to show that the energy function $\phi(V)$ must satisfy (7.13) we formulate first a slightly more general problem.

---

†In applying mathematics to science, one always assumes that the functions involved have as many derivatives as needed for the investigation. This optimistic expectation is indeed most often fulfilled. But to ease our mathematical conscience, we should say that our results are valid provided that our implicit assumptions are fulfilled.

Suppose there are three functions f, g and h (each twice differentiable) such that the system of equations

$$(7.14) \qquad m_1 f(V_1) + m_2 f(V_2) = m_1 f(W_1) + m_2 f(W_2)$$

$$(7.15) \qquad m_1 g(V_1) + m_2 g(V_2) = m_1 g(W_1) + m_2 g(W_2)$$

$$(7.16) \qquad m_1 h(V_1) + m_2 h(V_2) = m_1 h(W_1) + m_2 h(W_2)$$

has a unique non-trivial solution W_1, W_2 for arbitrarily given m_1, m_2, V_1, V_2. What can we say about the three functions? We shall show that they are linearly related; that is, they satisfy a relation

$$af(W) + bg(W) + kh(W) = d$$

with constants a, b, k, d, for all W. These functions will be identified later with ϕ and its first two derivatives.

Let us first simplify our notation. If we divide all three equations by m_1, we see that everything depends only on the ratio of masses

$$\lambda = \frac{m_2}{m_1}.$$

Next, we define a number triplet depending on V, which we write

$$(7.17) \qquad r(V) = \begin{pmatrix} f(V) \\ g(V) \\ h(V) \end{pmatrix}$$

and interpret as a vector in space, in analogy to our notation for vectors in the plane, discussed in Chapter 4. We can then contract equations (7.14)–(7.16) into the single vector equation

$$(7.18) \qquad r(V_1) + \lambda r(V_2) = r(W_1) + \lambda r(W_2).$$

We differentiate this equation with respect to V_1, holding V_2 and λ fixed, and find

$$(7.19) \qquad r'(V_1) = r'(W_1)\frac{\partial W_1}{\partial V_1} + \lambda r'(W_2)\frac{\partial W_2}{\partial V_1},$$

ENERGY

where we use the obvious notation

$$(7.17') \qquad r'(V) = \begin{pmatrix} f'(V) \\ g'(V) \\ h'(V) \end{pmatrix}.$$

Similarly, differentiation with respect to V_2 yields

$$(7.20) \qquad \lambda r'(V_2) = r'(W_1)\frac{\partial W_1}{\partial V_2} + \lambda r'(W_2)\frac{\partial W_2}{\partial V_2}.$$

It is always possible to find a vector

$$s = \begin{pmatrix} a \\ b \\ k \end{pmatrix}$$

such that for fixed V_1 and V_2 the two vector equations

$$(7.21) \qquad s \cdot r'(V_1) = af'(V_1) + bg'(V_1) + kh'(V_1) = 0$$

and

$$(7.22) \qquad s \cdot r'(V_2) = af'(V_2) + bg'(V_2) + kh'(V_2) = 0$$

are fulfilled. This follows immediately from algebra, since we have here two linear equations for the three numbers a, b, k. It has a geometric interpretation: s is the vector orthogonal to $r'(V_1)$ and $r'(V_2)$. Observe the important fact that s depends only on V_1 and V_2, not on λ.

Equations (7.19), (7.20) express $r'(V_1)$ and $r'(V_2)$ as linear combinations of $r'(W_1)$ and $r'(W_2)$. We assume these relations can be solved for $r'(W_1), r'(W_2)$ in terms of $r'(V_1), r'(V_2)$. It then follows from (7.21), (7.22) that s is orthogonal also to $r'(W_1), r'(W_2)$:

$$(7.23) \qquad s \cdot r'(W_1) = s \cdot r'(W_2) = 0.$$

Let us differentiate (7.18) with respect to λ, holding V_1 and V_2 fixed; we find

$$(7.24) \qquad r(V_2) - r(W_2) = r'(W_1)\frac{\partial W_1}{\partial \lambda} + \lambda r'(W_2)\frac{\partial W_2}{\partial \lambda}.$$

After scalar multiplication with s and use of eq. (7.23), we get

(7.25) $$s \cdot r(V_2) - s \cdot r(W_2) = 0.$$

Using the abbreviation $s \cdot r(V_2) = d$, we write (7.25) in the form

(7.25)' $$d = s \cdot r(W_2) = af(W_2) + bg(W_2) + kh(W_2).$$

Still holding V_1 and V_2 fixed, let us now change λ; W_1 and W_2 will range over some intervals, and for all W in the interval swept by W_2, (7.25') holds:

(7.26) $$s \cdot r(W) = af(W) + bg(W) + kh(W) = d.$$

Since s and d were completely determined by V_1 and V_2, we come to the conclusion that around each value W_2, there is a whole interval in which the three functions $f(W)$, $g(W)$ and $h(W)$ are linearly connected according to (7.26). It is not difficult to show (for a skillful mathematician) that a relation (7.26) holds for all values of W; we skip the technical details.

We have now shown that every admissible energy function $\phi(V)$ must satisfy a second order differential equation (7.13) with constant coefficients. Observe that $\phi(V) = \tfrac{1}{2}V^2$ satisfies it; indeed

$$\phi''(V) = 1,$$

as was to be expected. But there are more choices possible; we explore these in the next section.

7.4 More About Admissible Energy Functions

We now further restrict the class of energy functions by requiring them to be *even* functions, i.e., we require that for all V

$$\phi(-V) = \phi(V).$$

The physical basis of this requirement is that the energy of a body should depend on the speed with which it moves, but not on whether it is moving right or left (this is a symmetry principle).

If a function ϕ is even, it follows from the chain rule that its derivative $\phi'(V)$ is odd, i.e., $\phi'(-V) = -\phi'(V)$, and if a function is odd, its derivative is even. If we replace V by $-V$ in (7.13), we find the new equation

(7.13') $$a\phi''(V) - b\phi'(V) + k\phi(V) = d.$$

ENERGY

This is consistent with (7.13) for non-constant ϕ only if $b = 0$, and the equation for ϕ is of the form

(7.27) $$a\phi''(V) + k\phi(V) = d.$$

Certainly $a \neq 0$, otherwise the energy is constant.

The case $k = 0$ leads again to the solution $\phi(V) = (d/2a)V^2$, which we discussed earlier as the Newtonian choice.

If $k \neq 0$, it is easy to see that every solution of (7.27) is a sum of a solution of

(7.27') $$a\phi''(V) + k\phi(V) = 0$$

and a constant. We turn now to equation (7.27'). Let us first take the case that k/a is positive; then set

$$\frac{k}{a} = \gamma^2 > 0, \quad \gamma \text{ real}.$$

In this case equation (7.27') reads $\phi'' = -\gamma^2\phi$. Its solution is well known; it is of the form

(7.28) $$\phi(V) = A\sin(\gamma V + \beta),$$

where A and β are constants of integration. This $\phi(V)$ is an oscillating function of the speed and even takes negative values for some choices of V. We discard this possibility on physical grounds since we wish the energy to increase monotonically with the speed V.

Thus necessarily

$$\frac{k}{a} = -\alpha^2, \quad \alpha \text{ real}.$$

The solution of $\phi''(V) = \alpha^2\phi(V)$ is of the form $\phi(V) = Ae^{\alpha V} + Be^{-\alpha V}$. Since we demanded $\phi(-V) = \phi(V)$, we have the condition $A = B$, and so

(7.29) $$\phi(V) = A(e^{\alpha V} + e^{-\alpha V}).$$

7.5 Proof that $\phi(V)$ is Admissible

In trying to satisfy the requirement that the conservation laws (7.11) have a unique non-trivial solution, we were led to $A(e^{\alpha V} + e^{-\alpha V})$ as candidate for an admissible energy function. We now show by means of algebraic considerations that (7.29) does indeed satisfy this requirement.

Let us divide (7.11) by m_1; using the abbreviation previously introduced $\lambda = m_2/m_1$, we write (7.11) as

(7.30) $\quad \phi(V_1 - U) + \lambda\phi(V_2 - U) = \phi(W_1 - U) + \lambda\phi(W_2 - U).$

Inserting our energy function (7.29) into (7.30), we obtain

(7.31) $\quad A\big(e^{\alpha(V_1-U)} + e^{-\alpha(V_1-U)}\big) + \lambda A\big(e^{\alpha(V_2-U)} + e^{-\alpha(V_2-U)}\big)$

$\quad = A\big(e^{\alpha(W_1-U)} + e^{-\alpha(W_1-U)}\big) + \lambda A\big(e^{\alpha(W_2-U)} + e^{-\alpha(W_2-U)}\big).$

Each term contains either the factor $e^{-\alpha U}$ or $e^{\alpha U}$; for (7.31) to hold for all U, corresponding coefficients of these terms must be equal. After dividing (7.31) by A we equate these corresponding coefficients:

$$e^{\alpha V_1} + \lambda e^{\alpha V_2} = e^{\alpha W_1} + \lambda e^{\alpha W_2},$$

$$e^{-\alpha V_1} + \lambda e^{-\alpha V_2} = e^{-\alpha W_1} + \lambda e^{-\alpha W_2}.$$

We introduce the convenient notation

(7.32) $\quad e^{\alpha V_1} = x, \quad e^{\alpha V_2} = y, \quad e^{\alpha W_1} = \xi, \quad e^{\alpha W_2} = \eta;$

then these equations become

(7.33) $\quad\quad\quad\quad\quad x + \lambda y = \xi + \lambda\eta,$

$$\frac{1}{x} + \frac{\lambda}{y} = \frac{1}{\xi} + \frac{\lambda}{\eta}.$$

We rewrite them as

$$x - \xi + \lambda(y - \eta) = 0,$$

$$\frac{1}{x\xi}(x - \xi) + \frac{\lambda}{y\eta}(y - \eta) = 0.$$

Using the first equation to eliminate $\lambda(y - \eta)$ in the second, we obtain

$$\left(\frac{1}{x\xi} - \frac{1}{y\eta}\right)(x - \xi) = 0.$$

Since we are looking for non-trivial solutions, $W_1 \neq V_1$, i.e., $x \neq \xi$, see

(7.32), we divide by $x - \xi$ and conclude that

$$\frac{1}{x\xi} - \frac{1}{y\eta} = 0, \qquad \text{so} \qquad x\xi - y\eta = 0.$$

This and the first equation in (7.33) yield the linear system

$$\xi + \lambda\eta = x + \lambda y,$$
$$x\xi - y\eta = 0$$

for ξ, η; its solution is

(7.34) $$\xi = y\frac{x + \lambda y}{\lambda x + y}, \qquad \eta = x\frac{x + \lambda y}{\lambda x + y}.$$

Since x, y, λ are positive, so are ξ, η, and we can uniquely determine W_1, W_2 from (7.32). This solution is non-trivial, for if $V_1 \neq V_2$, then $x \neq y$, and by (7.34), $\xi \neq \eta$; therefore $W_1 \neq W_2$. This completes the verification that (7.29) is an admissible energy function.

7.6 Energy and Momentum

Our candidate $\phi(V) = A(e^{\alpha V} + e^{-\alpha V})$ for relativistic energy contains two parameters, A and α, whose values have to be determined. In the last section we saw that ϕ is an admissible energy function for all values of A and α. We need another principle in physics to determine A and α; this turns out to be momentum.

We shall use a method typical of research in physics. Suppose we have a theory that works well in a certain range of the variables, but fails if we push too far. We need a new, more general, theory with new functions and parameters. We then demand that the new theory almost coincide with the old in the domain where the old one worked well. This often guides us in choosing the new parameters appropriately. The importance of this method, called the *correspondence principle* of physics, was stressed by Niels Bohr, who applied it vigorously in his exploration of quantum theory.

Classical mechanics works very well for velocities v much smaller than the velocity c of light. Thus, when v/c is very small, we should demand that relativistic and Newtonian energy are nearly equal. Using (7.9) for the relativistic energy, we have

(7.35) $$E = m_0\phi(V) \approx \frac{m_0}{2}v^2 \qquad \text{for} \qquad v/c \ll 1.$$

Here m_0 is the rest mass (see Section 7.2) of the moving body. Now we recall the relation (7.8)

$$V = \frac{1}{2} \ln \frac{1 + v/c}{1 - v/c}$$

between ordinary and rescaled velocities. This relation is equivalent to

(7.36) $$e^{2V} = \frac{1 + v/c}{1 - v/c};$$

and, when v/c is very small, so is V. Using the Taylor expansion

$$\ln(1 + x) = x - \frac{x^2}{2} + \frac{x^3}{3} - \cdots,$$

we find that

(7.37) $\quad 2V = \ln(1 + v/c) - \ln(1 - v/c) = 2(v/c) + \frac{2}{3}(v/c)^3 + \cdots;$

and using the Taylor expansion

(7.38) $$e^x = 1 + \frac{x}{1!} + \frac{x^2}{2!} + \frac{x^3}{3!} + \cdots$$

we have, by virtue of (7.29),

$$\phi(V) = A(e^{\alpha V} + e^{-\alpha V}) = A(2 + \alpha^2 V^2 + \cdots)$$

$$= 2A + A\alpha^2 \left((v/c)^2 + \cdots \right).$$

Thus

$$\phi(V) - 2A = A\alpha^2 \frac{v^2}{c^2} + \cdots.$$

We recall from Section 7.4 that ϕ may be modified by adding a constant. Thus to satisfy (7.35) we stipulate that for small v/c

$$A\alpha^2 \frac{v^2}{c^2} \approx \frac{v^2}{2};$$

ENERGY

this necessitates the relation

(7.39) $$\frac{A\alpha^2}{c^2} = \frac{1}{2}$$

between the parameters A and α. We solve it for A:

$$A = c^2/2\alpha^2.$$

When we set this into the expression (7.29) for ϕ, we obtain

(7.40) $$E = m_0\phi(V) = \frac{m_0 c^2}{2\alpha^2}(e^{\alpha V} + e^{-\alpha V}),$$

an expression for the energy containing only one parameter, α.

To pin down the value of α, we apply the correspondence principle again. We differentiate the conservation law (7.11) with respect to U and obtain

(7.41)
$$m_1\phi'(V_1 - U) + m_2\phi'(V_2 - U) = m_1\phi'(W_1 - U) + m_2\phi'(W_2 - U).$$

This shows that $m\phi'$ is a quantity *preserved* in elastic collisions. Since in Newtonian mechanics the only preserved quantity other than mass and energy is momentum, we shall identify $m\phi'$, or rather a constant multiple of it, with *relativistic momentum*, and denote it by M. Using (7.29) for ϕ we obtain, after differentiating,

(7.42) $$M(V) = K(e^{\alpha V} - e^{-\alpha V}),$$

where K is a constant to be determined. Expanding the exponentials by (7.38), we find

$$e^{\alpha V} - e^{-\alpha V} = 2\alpha V + \tfrac{1}{3}(\alpha V)^3 + \cdots;$$

so for small V, we have $M(V) \approx 2K\alpha V$. Expanding V in powers of v/c by (7.37), we get, for small v/c

$$M(V) \approx 2K\alpha(v/c).$$

The Newtonian momentum, on the other hand, is

$$M_{\text{Newt}} = m_0 v;$$

so the correspondence principle dictates that

$$2K\alpha/c = m_0, \qquad \text{i.e.} \quad K = m_0 c/2\alpha.$$

Setting this into (7.42) yields

(7.42′) $$M(V) = \frac{cm_0}{2\alpha}(e^{\alpha V} - e^{-\alpha V}).$$

Another application of the correspondence principle is needed to determine α.

Suppose a force of magnitude F acts on a particle during a time interval Δt. How does this change the momentum and energy of the particle? According to the laws of Newton, the changes in classical momentum and energy are

(7.43) $$\Delta M_{\text{Newt}} = F\Delta t, \qquad \Delta E_{\text{Newt}} = F\Delta x,$$

where x is the distance traversed by the particle while the force is applied. For a particle moving with velocity v, $\Delta x \approx v\Delta t$.

We can eliminate F by forming the ratio of the change in energy to the change in momentum:

(7.44) $$\frac{\Delta E_{\text{Newt}}}{\Delta M_{\text{Newt}}} = \frac{\Delta t}{\Delta x} \approx v.$$

We now express the first order relativistic change in energy as the derivative of the energy function (7.40):

(7.45) $$\Delta E \approx \frac{dE}{dV}\Delta V = \frac{m_0 c^2}{2\alpha}(e^{\alpha V} - e^{-\alpha V})\Delta V.$$

Similarly, the relativistic change in momentum can be expressed by the derivative of the momentum function (7.42′)

(7.46) $$\Delta M \approx \frac{dM}{dV}\Delta V = \frac{m_0 c}{2}(e^{\alpha V} + e^{-\alpha V})\Delta V.$$

Here ΔV denotes the change in V due to the action of the force. The ratio of changes in relativistic energy to momentum is

(7.47) $$\frac{\Delta E}{\Delta M} = \frac{c}{\alpha}\frac{e^{\alpha V} - e^{-\alpha V}}{e^{\alpha V} + e^{-\alpha V}} = \frac{c}{\alpha}\tanh \alpha V.$$

ENERGY

The correspondence principle tells us that for small Δt, the Newtonian and relativistic changes in energy and momentum are very nearly equal. Therefore the limit of the ratios in the two cases, (7.44) and (7.47) must be equal:

(7.47)′ $$v = \frac{c}{\alpha}\tanh \alpha V.$$

By means of (7.36), we express v/c as function of e^V:

$$\frac{v}{c} = \frac{e^{2V}-1}{e^{2V}+1} = \frac{e^V - e^{-V}}{e^V + e^{-V}} = \tanh V.$$

According to (7.47)′,

$$\frac{v}{c} = \frac{1}{\alpha}\tanh \alpha V.$$

We conclude that

$$\alpha = 1.$$

It remains to express relativistic energy and momentum in terms of the ordinary velocity v instead of the rescaled velocity V. We use formulas (7.40) and (7.42′) with $\alpha = 1$:

(7.48) $$E = \frac{m_0 c^2}{2}(e^V + e^{-V})$$

(7.49) $$M = \frac{m_0 c}{2}(e^V - e^{-V}).$$

By (7.36)

$$e^V = \left(\frac{1+v/c}{1-v/c}\right)^{1/2}, \quad e^{-V} = \left(\frac{1-v/c}{1+v/c}\right)^{1/2};$$

their sum is

$$e^V + e^{-V} = \frac{1+(v/c) + 1-(v/c)}{(1-v^2/c^2)^{1/2}} = \frac{2}{\sqrt{1-v^2/c^2}},$$

and their difference is

$$e^V - e^{-V} = \frac{1+(v/c) - (1-(v/c))}{(1-v^2/c^2)^{1/2}} = \frac{2v/c}{\sqrt{1-v^2/c^2}}.$$

Setting these into (7.48) and (7.49) respectively gives

$$(7.48)' \qquad E = \frac{m_0 c^2}{\sqrt{1 - v^2/c^2}},$$

$$(7.49)' \qquad M = \frac{m_0 v}{\sqrt{1 - v^2/c^2}}.$$

7.7 The Dependence of Mass on Velocity

In both Newtonian and relativistic mechanics the observed energy and momentum of a particle depend on the velocity with which it moves relative to the observer. In relativistic mechanics, in contrast to Newtonian, also the mass of the particle depends on its velocity: the larger the velocity, the larger the mass.

Recall that the mass of a particle is a measure of its inertia; that is, the bigger the mass m, the harder it is to speed up the particle. Since in our new theory it is impossible to surpass, or even reach, the speed c of light, it is clear that the inertia of a particle (and hence also its mass) must tend to infinity as its speed v tends to c.

To determine what kind of a function m is of v, we imagine an experimental set-up in which we measure the mass of a fast-moving particle by the force it takes to *deflect* it from its path. We therefore look at both velocity and momentum as vectors v and M in a two-dimensional x, y-plane. Denoting the x- and y-components of velocity by v_x, v_y, we write

$$(7.50) \qquad v = (v_x, v_y);$$

then according to (7.49'),

$$(7.51) \qquad M = \frac{m_0 v}{\sqrt{1 - v^2/c^2}} = (M_x, M_y)$$

where $v = |v|$ is the speed of the particle, i.e.

$$(7.52) \qquad v = |v| = \sqrt{v_x^2 + v_y^2}.$$

ENERGY

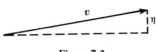

Figure 7.3

Suppose a particle travelling with speed v in the x-direction is acted upon by a force in the y-direction during a time interval Δt. This does not alter the x-component of relativistic velocity and changes its y-component from 0 to η, a small quantity of the order of Δt. The velocity of the particle now is

$$\boldsymbol{v} = (v, \eta),$$

see Figure 7.3, and its speed is

(7.53) $\qquad |\boldsymbol{v}| = \sqrt{v^2 + \eta^2} = v + \text{something small},$

where

(7.53') $\qquad \text{something small} \leq \text{constant} \cdot \eta^2.$

So

$$|\boldsymbol{v}| \approx v,$$

where $\approx$ is the relation of two quantities differing by less than (7.53'). Then the change in the y-component of momentum is due mainly to the change in the y-component of velocity. So we write, see (7.51),

(7.54) $\qquad \Delta M_y \approx \dfrac{m_0}{\sqrt{1 - v^2/c^2}} \eta.$

Next we compute the Newtonian change $(\Delta M_y)_{\text{Newt}}$ of the y-component of the momentum. Since the relativistic mass of a particle depends only on its speed $|\boldsymbol{v}|$, not on the direction of its velocity, and since $|\boldsymbol{v}| \approx v$, the mass changes by less than (7.53') during the interval Δt. Therefore the main Newtonian change of momentum is due to the change in the y-component of velocity:

(7.55) $\qquad (\Delta M_y)_{\text{Newt}} \approx m \, \Delta v_y.$

The y-component of velocity is η, so

(7.56) $$(\Delta M_y)_{\text{Newt}} \approx m\eta.$$

By the correspondence principle applied to the y-component of the change in M, $\Delta M_y \approx (\Delta M_y)_{\text{Newt}}$; so by (7.54), and (7.56) we find

$$\frac{m_0}{\sqrt{1 - v^2/c^2}} \eta \approx m\eta.$$

This shows that

(7.57) $$m = \frac{m_0}{\sqrt{1 - v^2/c^2}}.$$

Note that when $v = 0$, $m = m_0$.

We set this into (7.48'), (7.49') to obtain

(7.58) $$E = mc^2$$

(7.59) $$M = mv.$$

The imaginary experimental setup we used to determine mass as function of velocity is an abstraction of a real experiment carried out for the first time by Kaufmann in 1901. He observed electrons (discovered in 1897) travelling at high speed v, and he deflected them by a magnetic field perpendicular to their trajectories. According to Newtonian theory, force = mass × acceleration; so, the velocity η acquired by the electrons in the perpendicular direction is

(7.60) $$\eta = \frac{F \cdot \Delta t}{m},$$

where F is the magnitude of the magnetic force and Δt is the time during which the force is exerted, i.e. the time the electrons spend in the magnetic field. We have

(7.61) $$\Delta t = \frac{L}{v},$$

L being the length of the aparatus; see Figure 7.4. The angle of deflection θ satisfies

(7.62) $$\frac{\eta}{v} = \tan \theta.$$

ENERGY

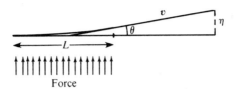

Figure 7.4

By means of (7.60), (7.61) and (7.62), m can be expressed as

$$(7.63) \qquad m = \frac{FL}{v^2 \tan \theta}.$$

F and L are parameters under the control of the experimenter. The speed v of the electron beam on which the experiment is carried out can be measured by a previous experiment. In the experiment under discussion the value of θ is measured. By (7.63) the value of m can then be determined.

It should be pointed out that the dependence of the mass of the electron on its velocity was not regarded as paradoxical in prerelativistic times. The apparent increase of mass was thought to be due to the electron pushing its way through the ether; several physicists, notably Abraham, Bucherer and Lorentz had quantitative theories. Only Lorentz took the outcome of the Michelson-Morley experiment into account by noting that the Fitzgerald-Lorentz contraction flattens the shape of electrons, and only his theory came up with Einstein's formula (7.57).

At Lorentz's request, Kaufmann repeated his experiment in 1906 to determine with greater accuracy the precise dependence of mass on velocity. His measurements indicated that (7.57) is wrong[†], much to Lorentz's disappointment. Einstein on the other hand was not at all distressed; so sure was he of the correctness of the theory of relativity that he immediately suspected an error in the experiment. Indeed, when Bucherer repeated it in 1908, he verified formula (7.57).

7.8 Energy and Matter

We saw earlier that equation (7.11), standing for the conservation of energy, determines the energy function only up to an additive constant. If we think of energy as kinetic energy, we would choose that constant so

---

[†] His data indicated a linear dependence of m on v.

that the energy of a particle at rest is zero, i.e.

$$\text{kinetic energy} = mc^2 - m_0c^2.$$

Einstein had the daring to take $E = mc^2$ seriously as expressing total energy, containing both kinetic energy and an energy equal to m_0c^2. This could be verified in processes where mass is annihilated and changed to energy.

By 1905, the Curies had isolated radium in sufficient quantities to observe that it gave off palpable amounts of energy which kept their sample several degrees above the temperature of its surroundings. The source of this energy was a profound puzzle. Einstein suggested that in the radioactive decay process, the mass of the decay products is less than the original mass of radium, and that this lost mass, according to the formula $E = m_0c^2$, is the source of the mysterious energy. In 1905 there was no way to measure quantitatively the amount of mass lost, but since then numerous experiments in nuclear physics have verified the equivalence (7.58) of mass and energy. The explosion of nuclear weapons is the most striking—and frightening—demonstration of the equivalence; the generation of nuclear energy is another. Although feared by some, the availability of nuclear energy may yet save mankind from fratricidal war waged for the control of rapidly shrinking resources of fossil fuel.

7.9 The Lorentz Transformation and the Momentum-Energy Vector

In both Newtonian and relativistic mechanics, the energy and momentum of a particle depend on the velocity of the particle; the nature of the dependence is, however, quite different in the two theories. In this section we shall compare the relativistic momentum and energy of a particle as perceived by two observers who move with constant velocity relative to each other. This is most easily done if we use rescaled velocities and exploit the properties of the hyperbolic functions defined in Section 6.2.

We have shown in Section 7.6 that the momentum M and the energy E of a particle moving with rescaled velocity V relative to an observer are given by equations (7.49) and (7.48), respectively. These formulas can be written more compactly in terms of the functions $\cosh V$ and $\sinh V$ defined by (6.17) and yield

(7.64) $\qquad M = m_0 c \sinh V, \qquad E = m_0 c^2 \cosh V.$

ENERGY

A second observer, moving with rescaled velocity U relative to the first, perceives the rescaled velocity of the same particle as $W = V - U$. He records its momentum and its energy as

$$\overline{M} = m_0 c \sinh W, \qquad \overline{E} = m_0 c^2 \cosh W.$$

To find how $\overline{M}, \overline{E}$ are related to M, E, we use the addition formulas (6.21) for the hyperbolic functions and equations (7.64):

$$\overline{M} = m_0 c \sinh(V - U) = m_0 c [\sinh V \cosh U - \cosh V \sinh U]$$

$$= (m_0 c \sinh V) \cosh U - (m_0 c^2 \cosh V) \frac{1}{c} \sinh U$$

$$= M \cosh U - E \frac{1}{c} \sinh U,$$

$$\overline{E} = m_0 c^2 \cosh(V - U) = m_0 c^2 [\cosh V \cosh U - \sinh V \sinh U]$$

$$= (m_0 c^2 \cosh V) \cosh U - (m_0 c \sinh V) c \sinh U$$

$$= E \cosh U - Mc \sinh U.$$

The resulting pair

$$\overline{M} = (\cosh U) M - \left(\frac{1}{c} \sinh U\right) E$$

$$\overline{E} = (-c \sinh U) M + (\cosh U) E$$

may be written in matrix notation as

(7.65) $$\begin{pmatrix} \overline{M} \\ \overline{E} \end{pmatrix} = \begin{pmatrix} \cosh U & -(1/c)\sinh U \\ -c \sinh U & \cosh U \end{pmatrix} \begin{pmatrix} M \\ E \end{pmatrix}.$$

Now we recall from Section 6.2 the definition (6.22) of a Lorentz matrix

$$L(U) = \begin{pmatrix} \cosh U & -c \sinh U \\ -(1/c)\sinh U & \cosh U \end{pmatrix}.$$

Replacing the energy E and $\overline{E}$ above by the modified dimensionless quantities

$$E^* = E/c^2, \qquad \overline{E}^* = \overline{E}/c^2$$

we may rewrite (7.65) in the form

$$\begin{pmatrix} \overline{M} \\ \overline{E}* \end{pmatrix} = L(U) \begin{pmatrix} M \\ E* \end{pmatrix}. \tag{7.66}$$

Thus the momentum-energy vector $\begin{pmatrix} M \\ E* \end{pmatrix}$ under a change of observer undergoes *the same linear transformation* as the space-time vector $\begin{pmatrix} x \\ t \end{pmatrix}$.

To write the momentum-energy vector in terms of the original velocity, we use relations (7.57), (7.58) and (7.59) and obtain

$$\begin{pmatrix} M \\ E* \end{pmatrix} = \begin{pmatrix} mv \\ m \end{pmatrix} \quad \text{where} \quad m = \frac{m_0}{\sqrt{1 - v^2/c^2}}. \tag{7.67}$$

The transformation law (7.66) for the momentum-energy vector leads to many applications and consequences that we cannot elaborate here. Our main aim was to show that our choice

$$\phi(V) = \cosh V$$

for the energy function has led to this elegant description of momentum and energy.

In developing scientific theories the beauty of formulas and such esthetic considerations as symmetries, analogies and correspondences play a larger part in guiding the researcher than is generally recognized. In such arguments the mathematical aspect plays an essential role.

7.10 Relativity in More than One Space Dimension

So far we have pretended that our universe has only one space dimension[†] and have developed the theory of relativity accordingly. Here we sketch with a very broad brush how to take one more space dimension into account. This is still one short of the three dimensions of the real world, but the considerations that will carry us from one to two readily enable us to add the third; the interested reader is encouraged to think about it.

We recall that the basic question of one-dimensional relativity theory is: what transformation relates the space-time coordinates (x, t) and

---

[†]The only exception is Section 7.7 where the use of an additional space dimension is essential.

ENERGY

$(\bar{x}, \bar{t})$ that two observers, moving with constant relative velocity with respect to each other, assign to the same event? Let us call these *relativistic* transformations.

The basic question of two-dimensional relativity is the same, except that now space-time vectors have three coordinates, (x, y, t) and $(\bar{x}, \bar{y}, \bar{t})$ respectively. The answer to the basic question is the same in three space-time dimensions as it is in two: The relativistic transformations are Lorentz transformations, i.e. linear transformations that preserve the quadratic form $x^2 + y^2 - c^2 t^2$:

$$(7.68) \qquad \bar{x}^2 + \bar{y}^2 - c\bar{t}^2 = x^2 + y^2 - c^2 t^2.$$

We can arrive at this answer either by repeating the analysis given for the case of two space-time dimensions, or by deducing it from the two-dimensional case; here we shall take the latter course.

Suppose that two observers in three dimensional space-time, moving relative to each other, have chosen their coordinate axes so that each appears to the other to be moving along the y, $\bar{y}$ axis, respectively. Then the y, t coordinates are related to the $\bar{y}$, $\bar{t}$ coordinates by a two-dimensional Lorentz transformation, while the measurement in the x-direction is unaffected by the motion, i.e. $\bar{x} = x$. In matrix language

$$(7.69) \qquad \begin{pmatrix} \bar{x} \\ \bar{y} \\ \bar{t} \end{pmatrix} = \begin{pmatrix} 1 & 0 & 0 \\ 0 & \cosh V & -c \sinh V \\ 0 & -\frac{1}{c} \sinh V & \cosh V \end{pmatrix} \begin{pmatrix} x \\ y \\ t \end{pmatrix}.$$

Suppose two observers are at rest with respect to each other. Then their measurement of time is the same, $\hat{t} = t$, and their space coordinates are related by a rotation that depends merely on how each chooses his coordinate axes. In matrix language

$$(7.70) \qquad \begin{pmatrix} \hat{x} \\ \hat{y} \\ \hat{t} \end{pmatrix} = \begin{pmatrix} \cos \theta & -\sin \theta & 0 \\ \sin \theta & \cos \theta & 0 \\ 0 & 0 & 1 \end{pmatrix} \begin{pmatrix} x \\ y \\ t \end{pmatrix}.$$

Both (7.69) and (7.70) are examples of relativistic transformations. Indeed, we can see directly that both are Lorentz transformations, for in (7.69) we have $\bar{x} = x$, $\bar{y}^2 - c^2 \bar{t}^2 = y^2 - c^2 t^2$ from which (7.68) follows; and for (7.70) we have $\hat{t} = t$, $\hat{x}^2 + \hat{y}^2 = x^2 + y^2$, from which (7.68) follows.

Let us denote the 3×3 matrices appearing in (7.69) and (7.70) by $L(V)$ and $R(\theta)$, respectively. Since, clearly, the composite of two relativistic transformations is again a relativistic transformation, it follows that any product

$$(7.71) \quad L(V_n)R(\theta_n) \cdots L(V_2)R(\theta_2)L(V_1)R(\theta_1)$$

is a relativistic transformation; in the language of algebra, relativistic transformations form a group. Just as clearly, Lorentz transformations form a group, for if two transformations preserve the quadratic form $x^2 + y^2 - c^2t^2$, so does their composite.

It is not hard to show that every direct (i.e. orientation-preserving) Lorentz transformation can be written in the product form (7.71); in fact at most three factors are needed.

Exercise 7.1. Calculate $R(\theta_2)L(V_1)R(\theta_1)$ with $\theta_1 = -\frac{\pi}{2}$, $\theta_2 = \frac{\pi}{2}$, V_1 arbitrary.

We shall not prove this fact, but merely comment on some properties of the Lorentz group.

In Section 6.2 of Chapter 6 we pointed out the analogy between the group of isometries and the Lorentz group; there is a similar analogy between the group of isometries of three dimensional space and the Lorentz group in three dimensional space-time. One of the points of similarities is a negative one: neither of the three-dimensional groups is commutative.

Exercise 7.2. (a) Show that in x, y, z-space rotation through θ about the x axis and rotation through ϕ about the y-axis generally do not commute.
(b) Show that $L(V)$ and $R(\theta)$ do not commute when $V \neq 0$, $\theta \neq 0$.

In Section 7.9 we have shown that the momentum-energy vectors in two different coordinates systems are related to each other by the same Lorentz transformation as the coordinates themselves. The same result holds in three dimensional space-time, if we define the momentum-energy vector, in analogy with (7.67), as

$$(7.72) \quad (mv_x, mv_y, m),$$

where (v_x, v_y) is the velocity vector of a moving point mass relative to an observer in the x, y, t system, $|v| = \sqrt{v_x^2 + v_y^2}$ its speed, and $m = m_0/\sqrt{1 - |v|^2/c^2}$ its relativistic mass. For, clearly the transformation law holds for (7.69), since that is the two-dimensional case augmented passively by an additional space dimension, and (7.70) is merely a Euclidean rotation. It follows then that the transformation law for the momentum-energy vector holds for all composites, i.e. for all Lorentz transformations.

We conclude this section by showing how three-dimensional Lorentz transformations can be used to determine anew the value of the parameter α appearing in formulas (7.40) and (7.42′). From (7.42′), we have $M = (cm_0/\alpha)\sinh \alpha V$; we set

$$\mathcal{M} = \alpha M = cm_0 \sinh \alpha V.$$

From (7.40), we have $E = (c^2 m_0/\alpha^2)\cosh \alpha V$; we set

$$\mathcal{E} = \left(\frac{\alpha}{c}\right)^2 E = m_0 \cosh \alpha V.$$

The analysis in Section 7.9 shows that if the vector $\begin{pmatrix} x \\ t \end{pmatrix}$ is transformed by

$$\begin{pmatrix} \cosh V & -c\sinh V \\ -\frac{1}{c}\sinh V & \cosh V \end{pmatrix},$$

then the momentum-energy vector $\begin{pmatrix} \mathcal{M} \\ \mathcal{E} \end{pmatrix}$ is transformed by

$$\begin{pmatrix} \cosh \alpha V & -c\sinh \alpha V \\ -\frac{1}{c}\sinh \alpha V & \cosh \alpha V \end{pmatrix}.$$

It follows then in three dimensional space-time that if x, y, t are transformed by $L(V)$ defined in (7.69), the correspondingly defined three-dimensional momentum-energy vector would be transformed by $L(\alpha V)$. This has the following curious consequence: Suppose that θ, ϕ, U and V are so chosen that

(7.73) $$L(V)R(\phi)L(U)R(\theta) = I,$$

i.e. so that rotation through θ followed by $L(U)$, followed by another rotation $R(\phi)$, followed by $L(V)$ leads back to the original set of space-time coordinates. Then this sequence of transformations must also leave the value of the momentum-energy vector unchanged. That means that if (7.73) holds, then also

(7.73)′ $$L(\alpha V)R(\phi)L(\alpha U)R(\theta) = I.$$

It is not hard to verify that this is the case only if $\alpha = 1$.

This derivation of $\alpha = 1$ is quite different from the one given in Section 7.6: neither Newtonian mechanics nor the correspondence principle was used. What was used instead is the invariance of the law of relativity when coordinates are rotated in the plane. This is another impressive demonstration of the power of symmetry.

7.11 Relativity and Electrodynamics

Our discussion of relativity so far was based entirely on reconciling the results of the Michelson-Morley experiment with our notions of position and time as recorded by observers moving with respect to each other. Historically, however, there was another source of the theory of relativity: the need to clear up a difficulty in electrodynamics. In fact, the title of Einstein's epoch-making paper is "On the electrodynamics of moving bodies". Let us see what the difficulty was, and how Einstein cleared it up.

What is electrodynamics? It is the theory, developed by Maxwell in the second half of the 19th Century, that describes the propagation of electromagnetic waves. The theory is in the form of a system of partial differential equations for two vector functions of space and time, one describing the strength of the electric field, the other the magnetic field. Later on we shall say a little—very little—about the nature of these equations. Here we merely remark that in his derivation, Maxwell made use of some properties of ether; however, Maxwell's equations have gloriously survived the demise of ether.

The laws of Newtonian mechanics are the same for two classical observers moving with constant velocity v with respect to each other. "Classical" here means that their coordinates are related to each other by

(7.74) $$\bar{x} = x + vt, \quad \bar{t} = t.$$

Since this invariance was known, in a rudimentary form, to Newton's great predecessor, Galileo, transformations of form (7.74) are called Galilean transformations. Maxwell's equations, on the other hand, are *not* the same when expressed in terms of coordinates that are related to each other by a Galilean transformation. This is the difficulty mentioned in the first paragraph.

In 1887 Lorentz had the idea that the reason Galilean transformations do not work for Maxwell's equations is that they take no account of the Lorentz-Fitzgerald contraction. Accordingly he wrote down a class of transformations that do take this contraction into account, Lorentz

ENERGY

transformations. Lo and behold, Maxwell's equations were invariant under Lorentz transformations.

Thus the difficulty could be formulated so: How come the equations of mechanics are invariant under Galilean transformations, while Maxwell's equations are invariant under Lorentz transformations? Here is where Einstein came in; he showed that the laws of Newton are definitely wrong. Einstein's proposed replacement, the theory of relativity, agrees well with Newton's laws at low relative speeds, and is invariant under Lorentz transformation. Thus is harmony restored between mechanics and electromagnetic theory.

It should be remarked that such harmony is philosophically essential for relativistic mechanics; for the propagation of light, a crucial concept in relativity, is an electromagnetic phenomenon.

We end this section by a series of exercises, suitable for those who have a nodding acquaintance with partial differentiation.

Maxwell's equations are, as we mentioned before, a system of equations for vector functions. It follows from them that each component f satisfies a single equation, the celebrated *wave equation*. For one space variable x this equation is

$$(7.75) \qquad \partial_t^2 f - c^2 \partial_x^2 f = 0.$$

Exercise 7.3. (a) Show that the wave equation is invariant under a Lorentz transformation. I.e., suppose that $f(x, t)$ satisfies (7.75); define $\bar{f}(x, t)$ by

$$(7.76) \qquad \bar{f}(x, t) = f(\bar{x}, \bar{t}),$$

where $\bar{x}, \bar{t}$ are functions of x, t defined by (5.28). Show that $\bar{f}$ also satisfies the wave equation (7.75)

(b) The three dimensional wave equation is

$$(7.77) \qquad \partial_t^2 f - c^2 \left(\partial_x^2 f + \partial_y^2 f \right) = 0.$$

Show that if f satisfies this equation, and if $\bar{f}(x, y, t)$ is defined by

$$\bar{f}(x, y, t) = f(\bar{x}, \bar{y}, \bar{t}),$$

where $\bar{x}, \bar{y}, \bar{t}$ are functions of x, y, t defined by (7.69), then $\bar{f}$ also satisfies the wave equation (7.77).

(c) Show that if $\hat{f}(x, y, t)$ is defined by

$$\hat{f}(x, y, t) = f(\hat{x}, \hat{y}, \hat{t}),$$

where $\hat{x}, \hat{y}, \hat{t}$ are functions of x, y, t defined by (7.70), then if f satisfies the wave equation (7.77), so does $\hat{f}$.

(d) Show that the three dimensional wave equation (7.77) is invariant under any Lorentz transformation.

Exercise 7.4. Show that all functions f of form

$$f(x, t) = h(x - ct) \quad \text{or} \quad f(x, t) = k(x + ct)$$

are solutions of the wave equation (7.75) for *arbitrary* functions h and k. These solutions represent waves travelling with speed c to the right or to the left, respectively.

Epilogue

Our aim has been to illustrate the power and elegance of mathematical reasoning in science with some examples ranging from the work of Archimedes to that of Einstein. We started with problems of the lever, the growth of populations and the mirror and ended up with problems of space travel and atomic energy. The early chapters dealt with subject matter for which we have a feeling, an everyday experience. We all have carried ladders; we all have the comforting assurance of bone borne intuition. In contrast, the latter chapters dealt with subject matter beyond our everyday experience, such as intergalactic travel at nearly the speed of light. Here our intuition has to be brain borne; and therein lies the real difficulty—and novelty—of our later chapters. As science and engineering move into "unaccustomed dimensions", that is, into the subatomic world of the very small and into the world of galaxies of the very large, our intuition fails and our only guide is that sixth sense, mathematics.

While many illustrations in this book dealt with questions of 20th Century science, the only mathematics we have used is algebra and calculus. Is it not surprising that calculus, developed nearly three centuries ago has the power to deal with subjects so modern? The philosopher Philipp Frank has compared the relation between mathematics and science to that between the sewing machine and the fashion industry: Fashions change fast, but the sewing machine serves them all.

This is not to say that there is no need for new mathematics nor that mathematics has stood still. On the contrary, it has developed tremendously; its increased power has dealt with a whole host of problems in modern science, but these must remain beyond the scope of this little book.

Further Reading. To the interested reader we recommend G. Pólya's "Mathematical Methods in Science", Volume 26 in this New Mathematical Library series. To the sophisticated reader we also recommend E. Mach's "The Science of Mechanics" published by the Open Court Publishing Co., Chicago,[†] a classic which has inspired generations of scientists, most outstanding among them, Einstein.

[†] The figures on our cover are based on illustrations reproduced in the 1960 edition of E. Mach's book, with the publisher's permission.

Index

Aaboe, Asger, 15
Abraham, 189
absorption factor, 30
addition
  formulas, 119, 159, 191
    for hyperbolic functions, 191
    of trigonometry, 119, 123, 157
  of velocities, 160
    relativistic, 152, 163, 166
additive function, 173
additive property, 112, 116
admissible energy functions, 178 ff.
affine maps, 112
algebra, 104, 131
  factor theorem of, 143
  of matrices, 119, 123
  of vectors, 107, 108
  rules of, 123
algebraization of geometry, 105
Allgaier-Hamppe, 1
analytic geometry, 104
angle
  of deflection, 188
  of incidence, 76, 77, 80, 81, 82
  of reflection, 76, 77, 80
  of refraction, 82
  of rotation, 145, 156
apothem, 64
Archimedes, 1, 3, 4, 6–9, 11, 12, 15, 16, 24, 59, 169, 199
  axiom of, 4
  law of the lever, 3, 8, 169
  "mechanical method" of, 15, 16
  symmetry proof, 80

arithmetic
  average, 34
  -geometric mean, 44
    inequality, 25, 69–74
  mean, 44, 69–74
  progression, 46
associative, 115, 118, 164
  binary operation, 164, 167
  property, 164
attractors, 52, 56
average (see arithmetic mean)

beam(s) of light, 94, 100
  interfering, 94
Bernoulli, James, 33
binary operation, 164, 165, 166
  infinitesimal, 167
binomial
  coefficients, 32
  theorem, 32
Bohr, Niels, 181
Bucherer, 189

calculus, 91, 99, 162
  differential, 87, 98
  of variations, 91
Cartesian coordinates, 110
Cassegrain, 101–103
Cauchy, 70
Cavalieri, 15
center of gravity, 12, 14, 19, 20
centroid, 10, 12, 14, 16

clock readings, 137
collision
 elastic, 174
 symmetric, 170, 174
commutative, 107, 110, 118, 119, 194
commute, 112, 194
component(s), 109, 110, 113, 117, 129
 transformation, 130
composite, 117, 123–126
composition
 of linear maps, 115
 of maps, 115, 118, 123
compound interest, 30, 31
compounding
 continuously, 33
 instantaneously, 33
congruence, 124, 125
conic section 12, 101–103
conics, 99
conservation
 law, 183
 of energy, 170, 172, 174, 189
 of momentum, 172
continuous
 compounding, 33
 function, 144, 173
 additive, 174
converge, 52, 56, 67
convergence, 32, 66, 68
coordinate(s), 105, 129
 axes, 129
 Cartesian, 110
 geometry, 108
 system, 130, 145
 transformation, 128
Copernicus, 136
corpuscular
 nature of light, 95
 theory of light, 91, 94
correspondence principle, 181, 183–185,
 188, 196
cosines, law of, 110
crooked lever, 16, 18–20
Curie, 190
Cusanus, 25, 59, 66
cycle
 stable, 55, 56
 steady, 52, 53
 of period 2, 52
cyclotron, 150

Darwin, 46
decimal fraction, 21
density, 22
Descartes, 66, 91, 108
differential calculus, 87, 98
differential equation(s), 44, 45, 58, 175,
 178
 partial, 196
differentiate, 99, 165, 177, 183
displacement, 105, 106, 124–126
distance, 124–127, 130
 preserving map, 124–126, 130, 144
distributive, 107, 110, 112
dot product, 107, 108, 110, 111, 113, 118,
 127, 128

e (Euler's number), 32, 33, 73, 74
Einstein, Albert, 1, 94, 132, 137, 139, 144,
 148, 150, 152, 189, 190, 196, 197, 199
Einstein's
 formula, 163, 189
 law for relativistic addition of
  velocities, 160, 161
 principle of relativity, 171
 space-time transformation
  problem, 132, 138
 special theory of relativity, 147
 theory of relativity, 94
 time contraction, 150
 transformation equations, 153
 transformation law, 155
elastic, 170
 collision, 173, 174
electrodynamics, 196
electromagnetic theory, 197
electron(s), 188, 189
 beam, 129
ellipse, 12, 82, 95–99, 101, 103
 foci of an, 96
 reflecting, 99
 tangent to an, 95
ellipsoidal cupola, 99
elliptical orbit, 82
energy, 169, 173, 174, 184–186, 189, 190
 and matter, 189
 and momentum, 181
 conservation, 170, 172, 174, 189
 function, 174–181, 184, 192
  admissible, 174, 178, 181
  Newtonian, 175

INDEX

kinetic, 170, 189, 190
  Newtonian, 181
  relativistic, 182
epidemic, 29
equilibrium, 3, 5–8, 10–12, 17–21, 72
  of the lever, 3, 80
equiprobable, 40
error(s), 34, 35, 43
  actual, 35
  combination of, 35
  Gauss' law of, 35, 42
  in combinations of observations, 43
  law of, 34, 36, 42
ether, 94, 163, 189, 196
Euclid, 61, 75–77, 91
  optics of, 75
  law of reflection, 91
euclidean motions, 124
even function, 39, 178
events, 38
  combined, 38
  independent, 38
experimentum crucis, 93, 94
exponential, 25
  function, 28, 34, 42
  law of growth, 25, 34

factor theorem of algebra, 143
Feigenbaum, Mitchell J., 55
Fermat, 16, 84, 85, 91, 93–95, 98
  law of refraction, 90, 91
  quickest path principle, 84 ff.
Fitzgerald-Lorentz contraction, 148, 149, 189
fixed point, 53–55, 125, 126
  of a function, 53, 54
  stable, 54
Fizeau, 163, 167, 168
focus (foci), 96, 99, 101, 103
force, 18, 22, 186–188
  magnetic, 188
  moment of a, 18
Frank, Philipp, 199
Fresnel, 163, 167, 168
function, 25, 82, 86, 140, 141
  additive, 173
  continuous, 144, 173
  differentiable, 165
  energy, 174 ff.
    admissible, 174 ff.

even, 39, 178
exponential, 28, 34, 42
  growth, 25
hyperbolic, 157, 191
linear, 142
momentum, 184
vector, 196, 197
functional equation, 25, 27, 34, 42, 45, 157, 159

Galilei, Galileo, 1, 6, 9, 18–20, 24, 84, 196
Gauss, Carl Friedrich, 34–36, 42, 43, 72
  law of errors, 42
  method of least squares, 44, 72
geometric mean, 69 ff.
geometric progression, 45
geometry
  analytic, 104
  coordinate, 108
gravitational field, 169
group, 155, 194
growth, 25, 57
  exponential law of, 25, 34
  factor, 49
  functions, 25 ff.
  law of, 28, 29, 42
  population, 29, 45, 47
  rate, 45

half-life, 30, 150, 151
  of muons, 150
  of radium, 30
Heiberg, 16
Heron of Alexandria, 76–78, 80, 81, 85, 91
  minimum principle, 85
  shortest path law of reflection, 90
hexagon, 64
  regular, 64, 67
homogeneity, 112
homogeneous map, 114
Huygens, Christian, 94, 95, 170–172
hyperbola, 12, 101, 103
hyperbolic
  cosine, 159
  functions, 157, 190, 191
  sine, 159
  tangent, 159, 161

identity, 115, 125, 126
  map, 115, 116
image, 105, 106, 112, 113, 117, 125, 127
  point, 112
  mirror-, 120, 121, 125
impact, 170, 173
incident ray, 76, 91
inclined plane, 18, 20–22, 35
independent events, 38
index of refraction, 90
inequality of arithmetic-geometric mean, 25, 69 ff.
inertia, 186
infinity, 99
intensity of light, 30
interest, 30, 31
  compound, 30, 31
  rate, 30
  simple, 30
interferometer, 134
invariance, 196
  laws of, 169
  of velocity of light, 147
  principle of, 172
inverse, 106, 115–117, 119, 122, 126, 154
  of a displacement, 106
  of a Lorentz matrix, 154
  of a one-to-one onto map, 115, 116
isometry, 124–128, 130, 144, 145, 156, 160

Kaufmann, 188
Kepler, 82, 83, 90, 91
  third law, 84
kinetic, 35
  energy, 170, 189, 190
  theory of gases, 35

law(s)
  of conservation of energy, 170
  of cosines, 110
  of errors, 34–36, 42
  of (in)sufficient reason, 3, 4
  of invariance, 169
  of the lever, 3, 10, 11, 14, 16, 169
  Verhulst's, 47, 48, 55, 58
least squares, 44
  Gauss' principle of, 72
Leibniz, 16, 91
lens, refracting, 100

lever, 3, 5–7, 11, 14, 19
  Archimedes' law of the, 3
  crooked, 16, 19
  weightless, 3
Lie, Sophus, 167, 168
light
  corpuscular theory of, 91, 94, 95
  interfering beams of, 94
  Newton's mechanistic theory of, 91
  propagation of, 81, 163
  rays, 75, 81, 94, 133, 134
    bending of, 81
  rectilinear propagation of, 91, 95
  speed of, 137, 152
limit, 15, 67, 69
  relation, 89
limiting value
  tending to a, 32
linear, 112, 114–116
  functions, 142, 143
  map, 113, 116, 121
  non-, 115
  transformations, 112, 118, 144, 192, 193
Lorentz, 153, 189, 196
  contraction, 148, 189, 196
  group, 194
  matrix, 153, 159, 160, 164, 191
  transformation, 190, 193, 195, 197, 198

Mach, Ernst, 199
magnetic field, 196
Malthus, 45, 46, 49
map(s), 112–115, 127
  composition of, 123
  distance-preserving, 124, 125, 144
  identity, 115
  linear, 112, 113, 116, 121
  non-linear, 118
  one-to-one, 115
  onto, 115
  origin-preserving, 126
mapping (see map)
mass, 183, 190
  and energy, 173
  relativistic, 194
  rest-, 173, 174
mathematical induction, 28, 70
matrix (matrices), 104, 113, 114, 117, 118, 122, 123, 128, 131, 152
  algebra of, 119, 123

INDEX

group of, 155
Lorentz, 153, 154, 159
multiplication, 117, 124, 154, 158
orthogonal, 127, 128, 130, 156
representation, 115, 121, 123, 126
matter, 189
Maxwell, 1, 25, 35, 36, 41, 42, 196
equations, 196, 197
law of errors, 34, 42
mean, 25
arithmetic, 44, 69–74
geometric, 69–74
measurement(s), 34, 35, 43, 44, 72
mechanics, 197
classical, 94, 181
equatons of, 197
laws of, 169
Newton's, 91
median of a triangle, 10, 11
method of exhaustion, 15
Michelson, A. A., 132, 133, 135, 136
Michelson-Morley experiment, 132 ff., 140, 147, 189, 196
minimum, 44, 79, 87
length, 78
path, 78
principle of Heron, 85
principle in optics, 102
mirror, 76, 91, 97, 98, 101, 122
elliptical, 97
image, 78, 80, 120, 121, 125
paraboloidal, 99, 100
plane, 76
reflecting hyperbolic, 102
molecules, 35
moment, 3, 8, 12, 14, 17, 18
of a force, 18
turning-, 3, 17
momentum, 172, 181, 183–186, 192
-energy vector, 190, 192, 194, 195
function, 184
Newtonian, 183
relativistic, 183, 190
Morley, E. W., 132, 136
motion(s)
euclidean, 124
of the earth, 133, 134
rigid, 124
multiplication of matrices, 118
muons, 150, 151

Napier, 32
natural logarithms, 32
natural selection, 46
Newton, Isaac, 1, 16, 44, 91–95, 101, 136, 170, 184, 197
mechanistic theory of light, 91
laws, 91
Newtonian
energy, 181
mechanics, 94, 183, 186, 196
momentum, 183
reflector, 103
non-linear, 115
mappings, 118
nuclear
energy, 190
physics, 190
weapons, 190

one-to-one, 115
onto, 115, 116
optics, 75 ff., 81, 94, 95, 102
orbit, 82
orthogonal, 110
matrices, 127, 128, 130, 131, 156
transformations, 131
unit vectors, 127
vectors, 110, 127

parabola, 12, 15, 16, 99
parabolic mirror, 99
paraboloidal
accumulator, 100
mirror, 103
reflector, 100
parallelogram property of displacements, 106
particle of light, 94
perpetual motion machine, 21
photons, 95
π (pie), 25, 59
Cusanus' calculation of, 59
recursive formula for, 59
Planck, Max, 94
planet, 82, 83, 84
Pólya, George, 199
population, 29, 45–47, 50, 57
competitive, 47
cooperative, 47
decreasing, 29

estimate, 57
  growth, 29, 45, 47
    over-, 46
  size, 47–49, 51, 53, 56
pre-image, 115
Principia Mathematica, 170
probability, 35–41, 43
  density, 41
proper time, 148–150
Ptolemy, 81, 82, 136
Pythagoras, 66
  theorem, 41, 88, 105, 135

quantum mechanics, 95
quantum theory, 94, 181
quickest path principle, 84 ff., 91

radar listening devices, 100
radiation, 99, 150
radioactive atoms, 29, 150
  decay, 29, 30, 190
radio telescope, 100
radium, 29, 30, 190
radio, 100
rainbow, 91
random process, 29
ray of light, 76, 91, 97, 99
  incident, 76, 91
  reflected, 76, 80, 93, 97–99
  refracted, 85
recursive equation, 25, 59, 61
reflection, 93, 98, 99, 120–128, 130
  angle of, 76, 77, 80
  law of, 77, 80, 95, 97
refracting lenses, 100
refraction, 81, 85, 93, 94
  index of, 90
  law of, 95
regular polygon, 25, 59, 62, 64, 65
relativity theory, 94, 132, 135, 163, 173,
    192, 196, 197
  special, 144, 168
  two-dimensional, 192
relativistic
  addition of velocities, 152 ff., 161, 163,
    166, 167
  energy, 175, 181
  mass, 187, 194
  mechanics, 186, 197
  transformation, 193, 194
  velocity, 187

rescaled velocities, 160, 163, 174, 182, 185,
    190
rigid motions, 124
rotation, 114, 120, 122, 124–130, 144, 156,
    196
  angle of, 145, 156
  euclidean, 195
  matrix, 164

shortest path, 76, 77, 79, 85, 90
Snell, 91
space dimension, 192
space-time dimensions, 193
space-time vectors, 153, 155
special theory of relativity, 144, 147, 168,
    169
Spencer, Herbert, 2
stable cycle, 55, 56
stable fixed point, 54, 55
standard clock, 139
standard measuring rod, 139
steady cycle, 52
steady state, 50–52, 58
Stevin, Simon, 1, 20, 21, 23, 24, 35
symmetry, 4, 79, 80, 120, 169, 171, 196
  principle, 143, 178
synchronized clocks, 137–139

Taylor expansion, 182
telescope
  Cassegrain's, 100
  Mount Palomar, 103
  radio, 100
  reflecting, 100
  refracting, 100
theory of evolution, 46
time, 132 ff.
  proper, 148, 150
  -scale, 150
transformation(s), 104, 118, 130
  composite, 115
  coordinate, 128
  distance-preserving, 124, 125, 144
  Galilean, 196, 197
  linear, 112, 118, 144, 192, 193
  Lorentz, 190, 193, 195, 197, 198
  origin-preserving, 130
  orthogonal, 130, 131
transpose, 128
trigonometry, 77, 87, 104, 119
  addition formulas of, 119, 123

uranium, 150

Verhulst, 46–48, 55, 57, 58
vector(s), 104–107, 109, 110, 112–116, 118, 127–129, 152
   addition, 106, 108
   algebra of, 107
   column, 108
   functions, 196, 197
   momentum-energy, 190, 192, 194, 195
   orthogonal, 127
   row, 108
   space-time, 153
   unit, 108, 121, 127, 128

velocity, 194
   zero, 106, 117, 128

velocity
   of the earth, 133, 135
   of light, 85, 94, 133, 136, 138, 139, 169
   invariance of, 147
   relativistic addition of, 152, 155
   rescaling, 156, 160, 163, 174, 183, 190

wave equation, 197, 198
wave motion, 94, 95
whispering points, 99